C. Böhm, T. Hopp, N. Ruiter, N. Duric (Eds.)

International Workshop on Medical Ultrasound Tomography

14.-15. Oct. 2019, Wayne State University, Detroit, Michigan, USA | Proceedings

International Workshop on Medical Ultrasound Tomography

14.-15. Oct. 2019, Wayne State University, Detroit, Michigan, USA | Proceedings

Edited by
Christian Böhm
Torsten Hopp
Nicole Ruiter
Neb Duric

KIT Scientific Publishing

Impressum

KIT Scientific Publishing

Karlsruher Institut für Technologie (KIT)
KIT Scientific Publishing
Straße am Forum 2
D-76131 Karlsruhe

KIT Scientific Publishing is a registered trademark
of Karlsruhe Institute of Technology.
Reprint using the book cover is not allowed.

www.ksp.kit.edu

Print on Demand 2021 – Gedruckt auf FSC-zertifiziertem Papier

ISBN 978-3-7315-1054-3
DOI 10.5445/KSP/1000124805

Preface

This volume represents the second in a series of proceedings from biannual international workshops held under the heading of Medical Ultrasound Tomography (MUST).
The inaugural workshop - MUST 2017 - was held in Speyer, Germany. Presented here are the proceedings from MUST 2019, a workshop that was held in Detroit, MI, USA on the grounds of Wayne State University.

We thank the Dean of the College of Engineering, Dr. Farshad Fotouhi, for his academic and financial support, and for providing access to the meeting venue. We also thank Dr. Juri Gelovani, the leader of the Molecular Imaging Program at the Karmanos Cancer Institute for use of program funds to support the logistical costs of the workshop. Without the support of Drs. Fotouhi and Gelovani, this meeting would not have been possible.

This workshop was notable for its historical context. A keynote speech delivered by Dr. Gary Glover kicked off the meeting. Dr. Glover was the first scientist to publish in-vivo sound speed images of the breast over 40 years ago!

Ultrasound Tomography is an emerging technology for medical imaging that is rapidly approaching clinical utility. Multiple research groups around the globe are engaged in research, spanning theory to practical clinical applications and commercialization. This two-day MUST workshop was held October 14-15, 2019. It was designed to be interactive and bring together the growing ultrasound tomography community, for the purpose of discussing and exchanging new ideas and research results, with each other and with researchers from related fields. The invited talks are listed below.

- Ancient Ultrasound Tomography and MRI Perspectives of Breast Cancer
  Gary Glover, Ph.D, Professor of Radiology and
  Professor of Neurosciences and Biophysics, Stanford University
- Ultrasound imaging with FWI: from breast to brain
  Lluis Guasch, Ph.D., Imperial College of London
- Transcranial Ultrasound Brain Imaging (TUBI)
  Solution for Point-of-Care Diagnosis of Traumatic Brain Injuries
  Roman Maev, Ph.D., University of Windsor
- Full-ring Photoacoustic Tomography:
  Light and Sound to Enhance Diagnosis of Breast Cancer
  Mohammad Mehrohammadi, Ph.D., Wayne State University
- The development of ultrasound tomography: A historical perspective
  Paul Carson, PhD., University of Michigan
- Clinical Implications of Screening Breast Ultrasound: Past, Present and Future
  Rachel Brem, M.D., George Washington University

The program of the two-day workshop included six invited talks, 21 oral presentations and 15 posters. The presentations covered the history and future trends of Ultrasound Tomography, the clinical motivation and status of breast imaging as well as commercial developments.

Dedicated discussion sessions with the audience covered the current challenges and future directions of ultrasound tomography. The conference was attended by 42 participants from 10 countries who actively contributed to a very successful workshop.

Oral and poster presenters were asked to submit a one-page abstract before the workshop, which was reviewed by the scientific committee. Authors of accepted abstracts as well as the invited speakers were asked to submit a full paper. This book comprises the written versions of most of the contributions presented during the workshop and thereby provides an overview of the state of the art in medical ultrasound tomography.

We would like to sincerely thank all the colleagues involved in the scientific and local committee for their commitment.

Neb Duric, Kierstin Fiscus and Juri Gelovani
Wayne State University, USA

Roman Maev
University of Windsor, Canada

Torsten Hopp and Nicole Ruiter
Karlsruhe Institute of Technology, Germany

Christian Boehm
ETH Zurich, Switzerland

Mark Anastasio
University of Illinois, USA

Detroit, July 2020

Special Acknowledgements

The members of the scientific and local organizing committees are indebted to Kierstin Fiscus for her untiring efforts in support of MUST 2019. She carried the bulk of the organizational burden before during and after the workshop. She facilitated the interactions between the organizing committees, and coordinated communications with individual attendees. Kierstin worked behind the scenes to coordinate hotels, caterers and restaurants in support of the workshop. Furthermore, she played a central role in gathering materials for these proceedings. Her professionalism facilitated all aspects of the organizational work and her good humor and cheerfulness made the whole process enjoyable. All in all, her support was vital to the success of MUST 2019.

Thank you Kierstin!

Workshop Organization

Scientific Committee

Neb Duric
Wayne State University, US
Delphinus Medical Technologies, US

Torsten Hopp
Karlsruhe Institute of Technology, Germany

Nicole V. Ruiter
Karlsruhe Institute of Technology, Germany

Christian Boehm
ETH Zurich, Switzerland

Mark Anastasio
University of Illinois, US

Local Organizing Committee

Juri Gelovani
Wayne State University, US

Neb Duric
Wayne State University, US
Delphinus Medical Technologies, US

Roman Maev
University of Windsor, Canada

Kierstin Fiscus
Wayne State University, US

Session Chairs

Nicole V. Ruiter
Karlsruhe Institute of Technology, Germany

Neb Duric
Wayne State University, US
Delphinus Medical Technologies, US

Peter Littrup
Delphinus Medical Technologies, US

Christian Boehm
ETH Zurich, Switzerland

Mohammad Mehrmohammadi
Wayne State University, US

Torsten Hopp
Karlsruhe Institute of Technology, Germany

Workshop Impressions

KIT
USCT
Refraction corrected transmission imaging
Bézier curves: first results with KIT 3D USCT
Motivation
Reconstruction results

Medical Ultrasound Tomography
WAYNE STATE
College of Engineering
Our Sponsors
Karmanos

of ultrasound computed tomography
based on distributed system

DISCUSSION

MEDRC

Design and performance of a Tonpilz transducer for low frequency medical ultrasound tomography

Table of Contents

Keynote and Invited Talks

Ancient Ultrasound Tomography and MRI Perspectives of Breast Cancer

Gary Glover, Ph.D.[1]

[1] *Stanford University, CA, USA*

Biography

Gary H. Glover received his PhD in Electrical Engineering from the University of Minnesota in 1969. He joined GE's Corporate Research & Development (CR&D) Labs in Schenectady, New York and studied solid state devices, computed ultrasound tomography and X-ray computed tomography until 1976, when he moved to GE's Medical Systems in Milwaukee to help transition fan-beam CT technology from CR&D. In 1980, he began the development of MRI as one of a team of five, and was thus instrumental in defining both the CT and MR products for GE. He joined Stanford's Radiology Department as Professor in 1990 and founded the Radiological Sciences Laboratory, dedicated to advancing biomedical imaging. His field of research is in MRI physics in general, and specifically in the development and application of functional MRI (fMRI) methods since 1993. His students' recent contributions include optimized techniques for acquisition and analysis of fMRI data, characterization of the dynamics of brain networks, development of real-time fMRI biofeedback methods, and multimodal neuroimaging using fMRI combined with EEG, fNIRS, fPET and functional MR Elastography, as well as with neuromodulatory transcranial electrical and magnetic stimulation.

He is a member of the US National Academy of Engineering and a Fellow of the American Institute for Medical and Biomedical Engineering (AIMBE) as well as the International Society for Magnetic Resonance in Medicine (ISMRM), for which he is also Past President. He holds a number of other awards including RSNA's Outstanding Researcher Award and ISMRM's Gold Medal, as well as Distinguished Investigator of the Academy of Radiology and the International Academy of Biomedical Engineers.

He has authored approximately 50 patents and published some 400 papers on his research.

Transcranial Ultrasound Brain Imaging (TUBI) Solution for Point-of-Care Diagnosis of Traumatic Brain Injuries

E. Malyarenko[1]

[1] *University of Windsor, Ontario, Canada, E-mail: em@tessonics.com*

Abstract

Portability, safety, speed, cost, user friendliness, and accuracy of the final diagnosis are the most important aspects of any clinical diagnostic equipment. Although dramatic progress has been made in the area of medical diagnostic imaging, contemporary equipment still often relies on ionizing radiation and is bulky, costly, and difficult to use. Hence the quest for miniaturization and safety continues, as well as the efforts to solve new challenging diagnostic problems and a few highly professional research groups working pretty hard in this field during last decade worldwide, including us.

In this presentation, the author will provide an overview of current R&D status in this field and in more details introduce our latest results in developing Transcranial Ultrasound Brain Imaging (TUBI) complex solution for accurate and rapid on-site diagnosis of certain types of Traumatic Brain Injuries, e.g. Brain Hematomas, without resorting to current gold standard imaging tools, such as computed tomography (CT) and magnetic resonance imaging (MRI).

This presentation will address a fundamental problem in ultrasonic imaging: the inverse problem, that is, the need to recover parameters/image of target objects hidden from external observation by a barrier (e.g. acoustically non-transparent obstacle like the human skull). Ultrasonic waves propagating through multilayer thick skull bone undergo multiple reflections, strong beam distortions, phase aberration, and get considerably attenuated. Target abnormalities behind the skull bone can be static (e.g. bone fragments and blood clots) and dynamic (e.g. vasculature blood flow disruptions such as hemorrhage and aneurysm), making the detection of head trauma even more difficult. Our solution involves new highly efficient real-time image formation algorithms, improving the transcranial image contrast for better pathology detection, and mathematical morphology-based interpretation for both static and dynamic target abnormalities that will ultimately produce high quality digital images of the human brain. Our final strategic goal is the clinical development and modification of the TUBI solution into an effective portable device for emergency personnel, sonographers, and physicians.

Clinical Implications of Screening Breast Ultrasound: Past, Present and Future

Rachel Brem, M.D.[1]

[1] *George Washington University School of Medicine & Health Sciences, D.C., USA*

Abstract

This presentation will discuss the importance and clinical implications of dense breast tissue, how screening breast ultrasound can improve the detection of breast cancer, the challenges of implementing screening breast ultrasound and what the future holds in terms of technological improvement in breast ultrasound in detecting mammographically occult breast cancer as well as the increasingly important role of ultrasound and ultrasound based technologies in individualized breast cancer risk assessment.

Biography

Dr. Rachel Brem is the Director of Breast Imaging and Intervention at The George Washington University Medical Center, Professor of Radiology and the Vice Chairman of the Department of Radiology. In 2015 Dr. Brem was named the Program Leader for Breast Cancer at the GW Cancer Center. Dr. Brem completed her undergraduate studies at Brandeis University followed by medical school at Columbia University where she graduated with honors. Dr. Brem completed both her Diagnostic Radiology Residency and Breast Imaging Fellowship at the John Hopkins Medical Institutions. Since completion of her training Dr. Brem was on the faulty at John Hopkins as the Director of Breast Imaging. Currently she is the Director of Breast Imaging and Interventional Center at the George Washington Medical Institution and Professor of Radiology. Dr. Brem's research and clinical interest includes Minimally Invasive Breast Biopsy as well as New Technologies for the Earlier Diagnosis of breast cancer focusing Whole Breast Ultrasound for women with dense breast tissue for improved breast cancer detection and risk assessment, Artificial Intellegence for breast cancer detection and Molecular Imaging of the Breast. She currently directs multi-institutional clinical trials evaluating novel approaches to improved diagnosis of breast cancer and is the recipient of numerous honors and awards including Best Cancer Doctors (Newsweek), Castle Connolly Best Doctors for Cancer and Top Doctor, Jewish Woman International's Ten Women to Watch, the prestigious fellowship in the American College of Radiology and the Society of Breast Imaging. She is an internationally recognized expert on Breast Cancer who is sought to speak on the subject nationally and internationally. Dr. Brem has extensively published, and is also committed to mentoring of students and residents. She is on the scientific advisory board of The Prevent Cancer Foundation as well as FORCE (Facing our risk of cancer, for women who are BR CA positive) and is a member of the Board of the Katzen Cancer Research Center. She currently serves on the Board of Directors of Delphinus Medical Technologies and Dilon Technologies and served on the Board of Directors of iCAD, inc from 2002-2019.

Dr. Brem is the Director of the George Washington University Mobile Mammography program (Mammovan), a mobile van that brings mammography to underserved communities to optimize the care of all women.

Dr. Brem was part of the formation of the Brem Foundation to Defeat Breast Cancer in 2004 with a group of committed patients and friends who have worked diligently to support translational research as well as education and clinical care in the underserved community. The Brem Foundation has recently been instrumental in passing dense breast legislation in the District of Columbia which has the highest death rate from cancer nationally.

A History of US Transmission Tomography Emphasizing Some Approaches Out of the Mainstream

P. L. Carson[1,2]

[1] *Department of Radiology, Unversity of Michigan, Ann Arbor, Michigan, USA*
E-mail: pcarson@umich.edu
[2] *Department of Biomedical Engineering, Unversity of Michigan, Ann Arbor, Michigan, USA*

Abstract

As head of the second group to present on ultrasound transmission tomography (UTT), I switched from emphasizing speed of sound (SOS) imaging to attenuation and pulse echo imaging. Confocal, 19 mm diameter transducers in translate-rotate geometry performed better in those roles than did small diameter transducers used in the Mayo Clinic and General Electric efforts designed primarily for SOS imaging. While bent ray SOS imaging was being explored in the mainstream, phase insensitive receivers were considered for less edge enhancement and other artifacts in attenuation imaging. This simpler approach to quantitative attenuation imaging has only recently been revived for medical imaging. Other body parts than breast were considered, as well as imaging of other tissue characteristics. In the mid 80's, clinical trials of commercial automated breast imaging systems were judged by radiology leaders to be unsuccessful and research funding of quantitative imaging and "over-sold" tissue characterization went down with those systems. S. Johnson's Techniscan worked consistently through two down decades and UCT improved considerably with good bent-ray and then full wave migration or inversion imaging. Acquired by QT Ultrasound, the 3D full wave reconstructions have now been performed with the system. Competition from the simpler, but well sampled ring array of Delphinus Medical Technologies has moved UTT close to clinical acceptance. We began work on limited angle transmission tomography in the mammographic geometry, leading to use of pulse echo information to fill in missing data. The concept of bulk attenuation coefficient was introduced to minimize domination of attenuation images by losses at major boundaries delineated in pulse echo and transmission modes. With less than full apertures, the distinction between transmission and pulse echo imaging becomes less distinct. Tomography of bulk tissue properties by pulse echo systems is again worth consideration, as is combination of transmission data with other ultrasound modes and thermoacoustic imaging for increased image quality and information.

Keywords: Medical Imaging, Review, Quantitative Acoustics

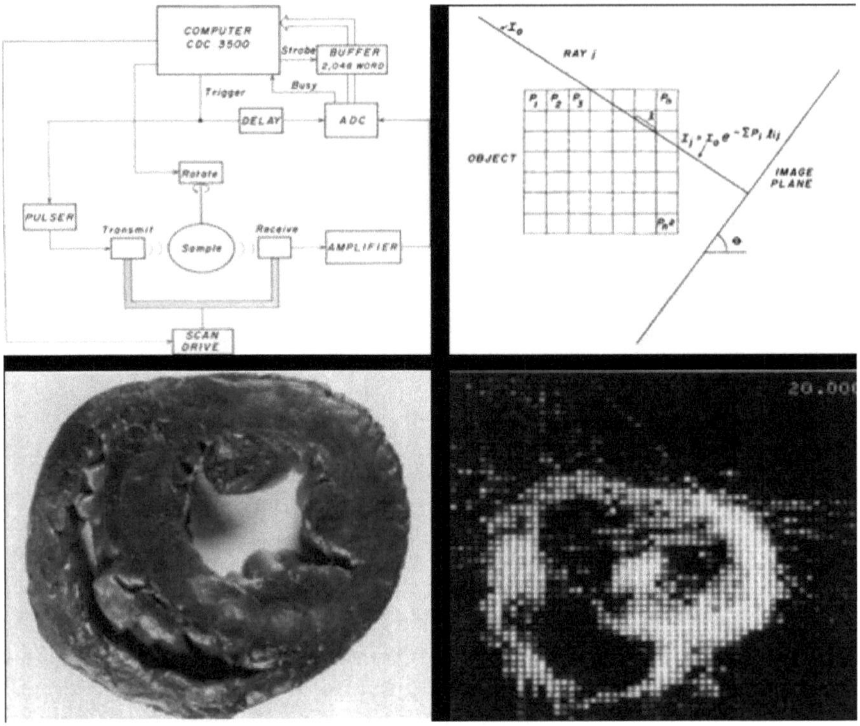

Figure 1: Clockwise from top left: From first UTT paper: Diagram of experimental system; schematic of analytic reconstruction technique; Algebraic reconstruction of acoustic "absorptions" within that approximate cross section; photo of transverse cross section through canine heart.

1 Earliest UTT

The term Ultrasonic CT or UCT is used with systems detecting the transmitted waves as is an x-ray CT. Taken literally, ultrasound computed tomography could be applied to the earliest pulse echo ultrasound images, if you don't limit the term to digital computing. I prefer ultrasound transmission tomography (UTT) for the transmission modes, ultrasound reflection tomography (URT) for the reflection modes and UCT in general. All borrowed images are adapted with permission from the adjacent references.

Greenleaf, Johnson and colleagues at the Mayo Research Foundation presented the first UTT constructions in July, 1973 (Figure 1) [1]. They modified a rectilinear radioisotope scanner for the rotate translate geometry with an 8 bit transient recorder and a 5 MHz transmitting and a 2.25 MHz receiving transducer. Reconstruction with the Algebraic reconstruction technique (ART) [2] produced ultrasound absorption images that were soon recognized as images of the broadband attenuation coefficient.

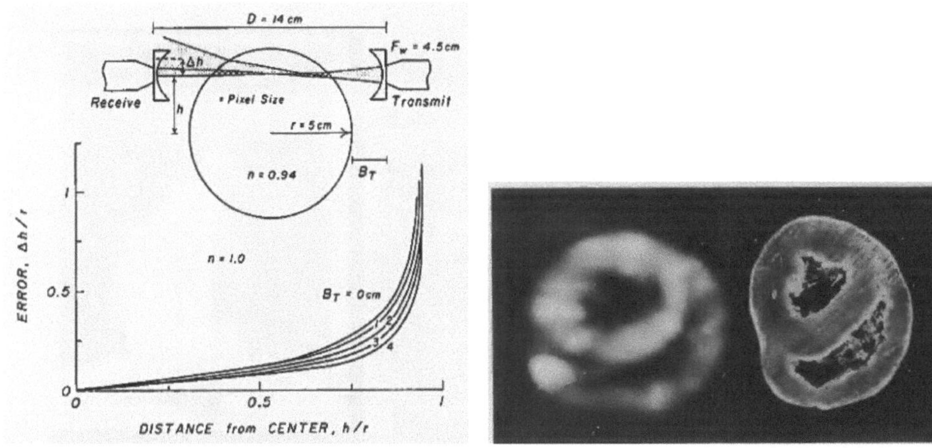

Figure 2: Left: Diagram of transmission tomography of the refactive index and plot of the deviation of the beam at the receiver, $\Delta h/r$, as a function of the fractional distance to the edge of imaged refracting cylinder [4]. Right: Image of the refractive index of a section through a heart, in vitro (left). A photograph of the approximate section of the heart (right).

The first time of flight images (Figure 2) were presented in 1975, showing strong promise for UTT [3]. Gary Glover at GE Global Research produced a comparable SOS imaging system ready for human studies, with results discussed later with other human studies.

With the help of the head of medical physics at the University of Colorado, William Hendee, I and, particularly, a medical student Gary Thieme, and others, built a treatment planning CT scanner using a Co^{6}0 source. I had just started medical ultrasound research and the extension to ultrasonic CT seemed possible. We started with Howry and Holmes' translate/rotate pulse echo water tank scanner from the 1950's, unaware of Greenleaf and Johnson's work until their presentation at the Seminar on Ultrasonic Tissue Characterization in May, 1975 [4]. As their best scanner was optimized for speed of sound (SOS) imaging, we began emphasizing attenuation imaging. The confocal, 19 mm diameter transducers, were expected to be better for spatial resolution in pulse echo and attenuation imaging of the breast, with less attenuation artifact in the latter [5]. Also for originality, we chose to explore the possibility of imaging through bone. Perry and Altschuler at NCAR in Boulder ran on one of the fastest commercial mainframes, a mathematically rigorous least-mean-squares method, similar in philosophy to that described by Marr's polynomial solution of an orthogonal set of Radon Transforms [6]. They achieved a bit more fidelity [5] than our filtered backprojection reconstruction shown in Fig. 3. Images of a lamb shank were of similar quality.

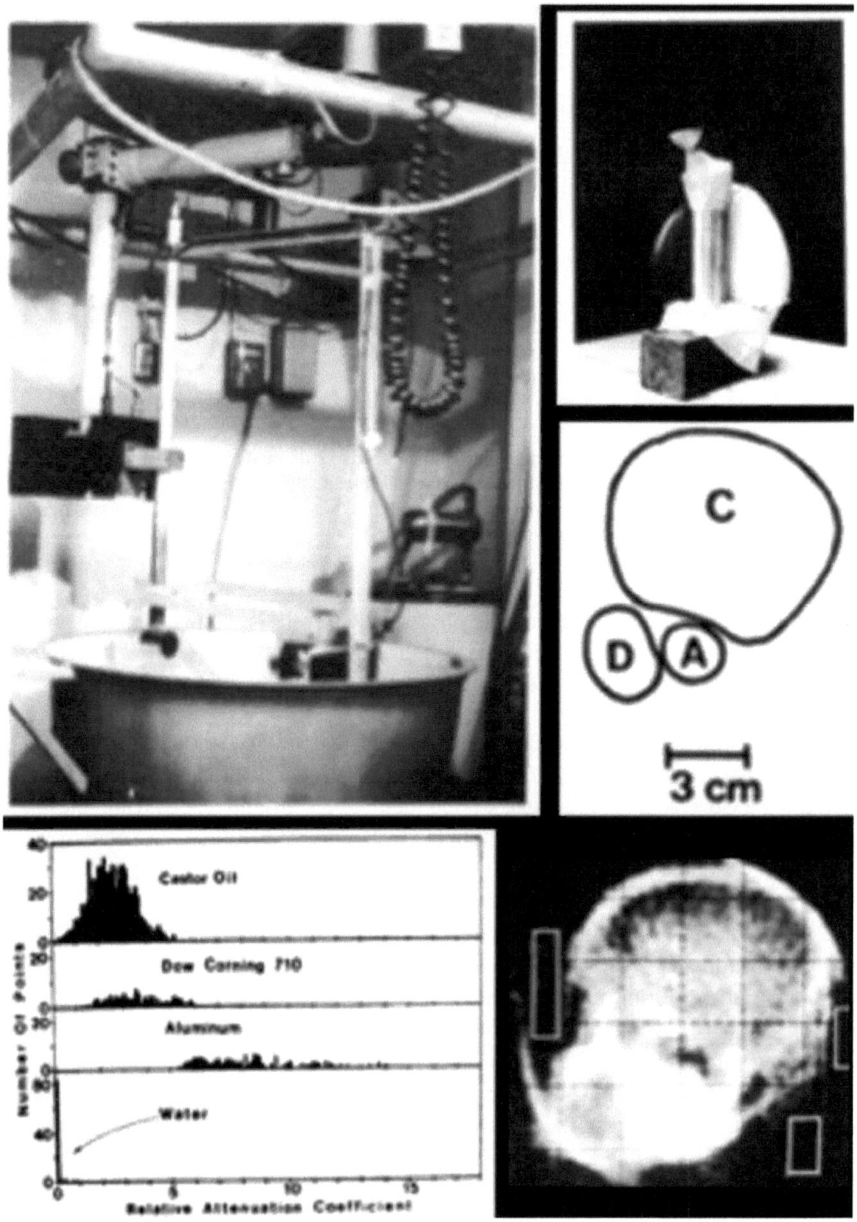

Figure 3: Clockwise from top left: (a) Water-path scanning mechanism adapted to UTT imaging; (b) Phantom comprised of a balloon of castor oil, one of Dow Corning 710 silicone oil and an aluminum rod; (c) Sketch of cross section of the phantom; (d) Broadband attenuation coefficient image through approximately the same cross section. Rectangular ROIs are show for water quantification; (e) We got semiquantitative information as shown in the histograms.

2 Reducing Effects in Attenuation Images of Diffraction at Tissue Boundaries

Avinash Kak at Purdue and a series of graduate students contributed many innovations over a decade. His earliest work on using just signal processing to reduce artifacts in the attenuation images were not very successful [7]. However, an earlier publication aimed at imaging the attenuation coefficient of tissue types rather than heavily edge-enhancing refraction/diffraction effects at borders [8,9], showed good paths forward (Fig. 4). The broadband correction in Fig. 4d) seems to do most poorly in that regard, showing the edge effects with high fidelity.

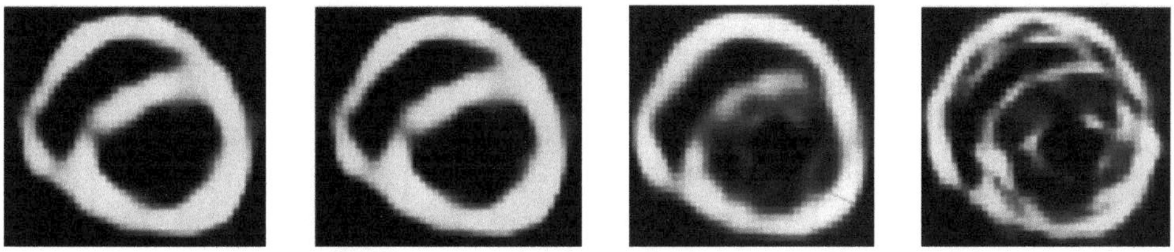

Figure 4: (a) Center frequency shift by Gaussian fit; (b) Energy ratio in lower/upper frequency windows; (c) Narrow band assumption; (d) Broadband correction.

The student Azimi worked with Kak on multiple scattering effects in diffraction tomography [10], a method of solving the wave equation with weak scattering assumptions, discussed later. In Fig. 5, with a weak scattering approximation (left) you get beautiful images of 2, 6λ cylinders, but if you include the presence of multiple scattering [11], you end up with the substantial artifacts (Fig. 5b), even for cylinders with breast tissue-like, 1.05, index of refraction. Anderson and Kak introduced the long popular SART algorithm with application to SOS UTT [12]. They state "we compute the individual correction terms..." for beam profile and length of path through a pixel. "The terms are then saved until all rays in that view have been considered. The average correction to each picture element is computed and added to form the updated image". They note that the method is applicable for curved ray paths in US and microwave tomography.

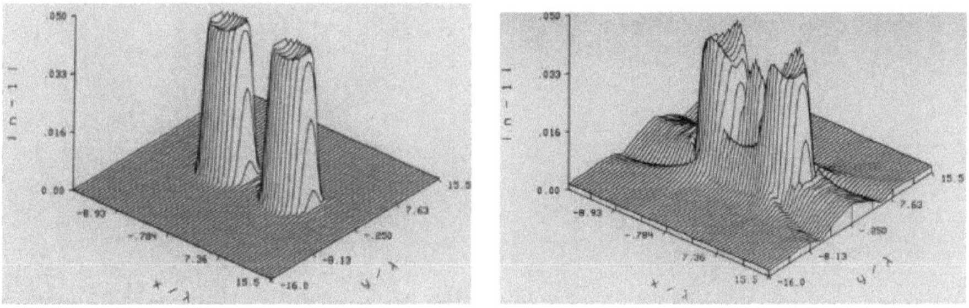

Figure 5: (a) Weak scattering using Born or Rytov approximation; (b) Doubly scattered fields included in the projection data.

Fig. 6 a)-c) shows reconstructions of the numeric Shepp-Logan Phantom. The single iteration of Art shows a very course noise from amplification of statistical fluctuations 6a). The reconstruction is much smoother and accurate using a single iteration of simultaneous ART 6b). Convolution back projection, however, gave an image 6c) very similar to that of the numerical phantom. At least both of the art methods used a longitudinal Hamming window, meaning along the path of the ultrasound ray. Zero weighting was at the boundaries of the Reconstructed overall object 6d). A.J. Devaney and others made early contributions to this approach [13-16].

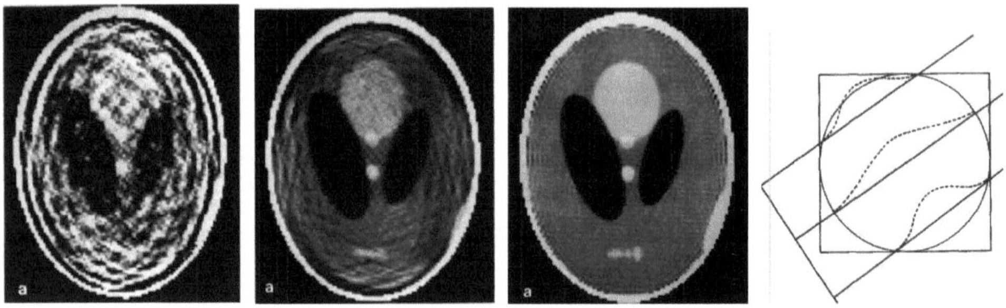

Figure 6: (a) Single iteration ART; (b) Single iteration SART; (c) Convolution backprojection; (d) Longitudinal Hamming window.

Using phase insensitive receivers to minimize phase cancellation artifacts and resulting edge enhancement in attenuation imaging was pioneered by Jim Miller and his students and post-docs particularly Klepper [17]. They produced a very prescient analysis of the various mechanisms for artifacts in attenuation images with transmission tomography and ranked them according to their probable importance in the reconstructed images. These and their proposed solutions for the varios types of artifacts are summarized well in Table 1 from their publication.

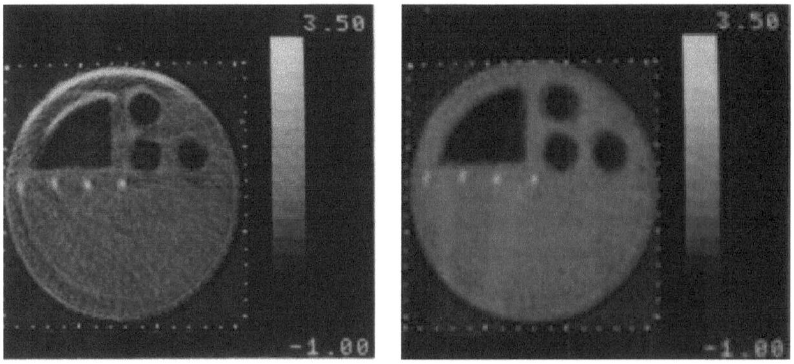

Figure 7: (a) Attenuation image, phase sensitive, focused, 19 mm piezoelectric receiver; (b) Phase-insensitive processing, with simulated linear array of 40 point-like elements.

Chenevert et al. showed that phase insensitive processing with point-like elements in a simulated 40 element linear array could eliminate many attenuation artifacts in a cylindrical geometry [18]. This demonstration was accomplished, for each transmitter position of a 19 mm diameter, single element transducer, by scanning a small needle hydrophone across the image plane path of the transmitted beam. The image in b) is quite good reconstruction of this phantom, which was designed to be difficult, given the straight edges of the upper left quadrant insert.

Modern phase insensitive Imaging is being done [19], achieving less edge enhancement of most of objects when the PVDF detectors are used with phase insensitive electronics. PVDF polymer detectors are not quite as sensitive as PZT, particularly on transmit. For reception, however, they are pretty comparable Zeqiri et al. are getting some interesting results with large detectors.

3 Other Algorithms

A classification of UTT reconstruction algorithms was published in 2013 [20] as:

- Ray-Based Backprojection Tomography
  - Straight Ray
  - ART, Filtered Back Projection
  - Bent Ray
- Diffraction Tomography
  - Weak Scattering
  - Distorted Wave Born Analysis
  - Inverse Scatter Tomography (IST)

3.1 Diffraction Tomography

In diffraction tomography, weak scattering is usually assumed in the form of the Born or Rytov approximation. An approximate solution to the wave equation was given by Mueller et al. [21] as: Write the function $F(r)$, where k is the wave number $2\pi/\lambda$ and c and c_0 are the local and surrounding medium speeds of sound, respectively.

"$F(\bar{r}) = k^2 \left(1 - \frac{c_0^2}{c^2(\bar{r})}\right)$. (1) ... write the wave equation as $\left(\nabla^2 + k^2\right) \psi = F(\bar{r})\psi$. (2)", where ψ is the particle velocity potential. "By definition the function $F(\bar{r})$ is zero for $|\bar{r}| > a$, the radius of the assumed circular object. We shall further assume that its mean is zero. This can always be achieved by appropriate choice of c_O. Eq. (2) can be transformed ... by introducing

$$u = \ln\psi \ (3) \quad \text{and substituting into (1)} \quad \nabla^2 u + (\nabla u)^2 + k^2 = F(u). \ (4)$$

If one considers $F(\bar{r})$ as a small perturbation one can formally solve (2) and (4) by a perturbation method. The first order approximation obtained by this method from (2) is known as Born's approximation ... and ... from (4) as Rytov's approximation ..."

Allowing for somewhat stronger interactions, the distorted wave Born approximation offers some improvement over simple Born or Rytov approximations [22]. A much more in-depth review of reconstruction algorithms is included in [23].

Estimated importance	1	2	3	4	5	6
Mechanism	Phase cancellation	Non-unity transmission coefficients	Frequency dependence of transmitted beamwidth	Tissue anisotropy	Loss of beam at receiver due to refraction	Path error due to refraction
Manifestation	Random over- estimation of attenuation yields only qualitative noisy image	Severe edge error, only qualitative image degradation	Significant edge artifact and image noise	Anisotropic regions reconstruct to consistent intermediate value, adjacent isotropic regions incorrectly estimated	Significant edge artifact	Geometry-dependent overall image
Correction	Intensity sensitive detection	Reconstruct the slope of the attenuation with respect to frequency	Spatially filter projections at each frequency, or use variable aperture transmitter	No known method	Wide aperture receiver combined with slope	Not addressed in the present work

Table 1: Mechanisms for artifacts in attenuation images with transmission tomography, ranked according to their probable importance in the reconstructed images.

4 Human Studies

The first human study of UTT was reported by Glover at GE [24]. The results in Table 2 show a trend of increasing speed of sound with expected connective tissue content. Later data shows some fibroadenomas with Cancer-like speeds of sound.

The first human study of SOS and relative attenuation coefficient, μ in numerous women [25] showed modestly promising results from imaging of 150 women with a total of 30 breast cancers from which 19 were analyzed in Fig. 8b. Attenuation was not a strong discriminant as might be expected from the image in Fig. 8a.

Results–Summary	CTU Nos
Benign	
Fluidic cyst	0–5
Fibroadenoma	20–28
Cystic Mastitis	20–30
Malignant	
Ductal, lobular carcinoma	40–55
Medullary, inflammatory	
Carcinoma, skin thickening	40–60
Asymptomatic young mammary parenchyma	8–60

Reference: 10 CTU Nos = 1% velocity change relative to water. Mammary fat = −60, Water = 0.

Table 2: Study on 16 patients with breast lesions and 4, young asymptomatic women [24]. Reprinted with permission.

This publication was followed by results on 78 women, including 40 cancers, with complete data from 175 scans [26]. They concluded that "Images of the local speed of ultrasound show high values in regions of muscle, breast parenchyma, medullary carcinoma and connective tissue and show low values in regions of fat. Images of acoustic attenuation show high values in regions of connective tissue and borders of scirrhus carcinoma and low values in regions of fat." If UTT of the relatively rare medullary carcinoma does indeed contribute to its B-mode discrimination from benign solid masses such as fibroadenoma, that might allow noninvasive diagnosis to avoid biopsying many, subsequently benign, solid masses.

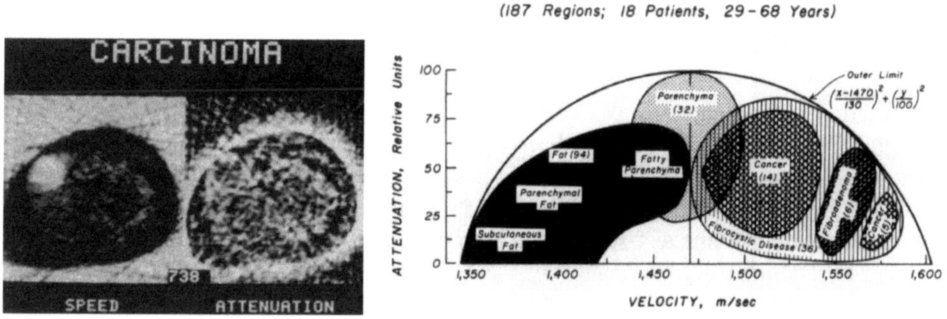

Figure 8: (a) Images of a carcinoma in upper left; (b) Attenuation and SOS plot for relevant tissues in breast with a mass.

In 40 participants with unequivocal lesions, from the 95 studied, the differences between SOS and μ in the mass and their values in the remaining central mammary tissues provided the discrimination in Fig. 9a [27]. These results were similar to those of the two earlier human studies that utilized the raw SOS and μ of the mass [25,26]. Actual SOS and μ ranges were very similar to those earlier studies, suggesting that normalization relative to,

Interpretation Modality	Sensitivity (%)	Specificity (%)	Positive Predictive Value (%)
Visual	62.5	63.2	64.1*
Computer-aided	82.5	60.5	68.8†
Mammography	92.5	34.2	59.7†

* $P < 0.05$. † $P < 0.01$. [redacted]

Table 3: Reprinted with permission from [26]. The increase in specificity and positive predictive value over mamography might be expected just from doing B-mode ultrasound.

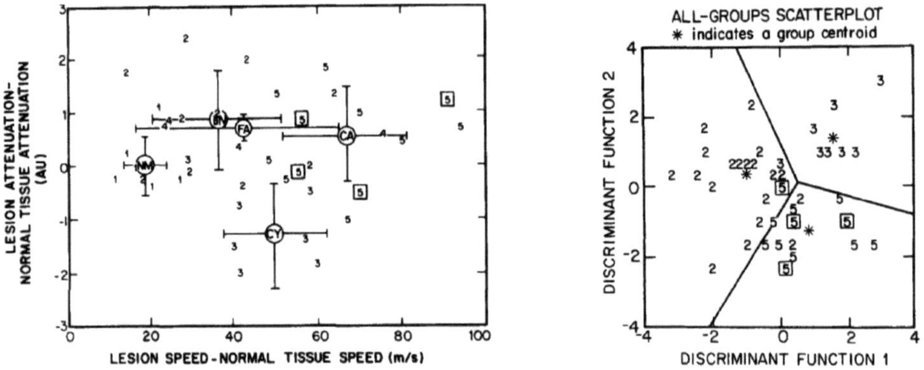

Figure 9: Results from [27]. Discrimination of three mass types by two linear discriminate functions of four tissues in the involved breast. The three masses are Benign solid - label 2, cystic - 3, and cancer - 5.

or inclusion of tissues other than the mass might not be necessary. This question was not clearly answered in this small study, but might be considered further with quantitative 3D imaging. Comparison with the contralateral breast for anatomy and tissue properties is certainly used in visual diagnoses. When SOS and μ in ten tissue regions of the participants' two breasts were evaluated by linear discriminate analysis, one nonmass tissue region contributed slightly to the discrimination. In order of importance in the discrimination of 3 key mass types were the tissue regions: a) the SOS in the mass including all of the border region seen in the attenuation image, b) SOS in the normal parenchyma of the involved breast, c) μ in the mass as defined on SOS image, i.e., usually not including the enhanced edge in the μ image, d) μ in the entire involved breast (Fig. 9b).

A good, recent plot of μ as a function of SOS, Fig. 10, is from a study with a commercial prototype scanner, now commercially available in an upgraded form. The images [28] are much better than previously. There is good separation between cancers in the red rectangles and other tissues, giving some hope for UTT as a screening, diagnostic and prognostic tool. This hope is particularly true if done in combination with pulse echo or other advanced reflection mode imaging comparable to current clinical ultrasound that provides better screening in dense breasts than digital mammography [29]. Despite diffraction artifacts providing edge enhancement in the attenuation images, μ seems to be a bit better discriminator than with earlier systems. Cysts, green circles, and compex cysts, blue circles, are well separated by μ. The general trends are still similar those of the earlier systems and investigators.

With the Delphinus system, substantial lesion characterization analysis was performed on reflection, SOS and μ images of 107 benign masses and 31 cancers [30]. Lesion boundaries and reflectivity were added tot he usual UTT measures. PPV was 91% for a sensitivity of 97%.

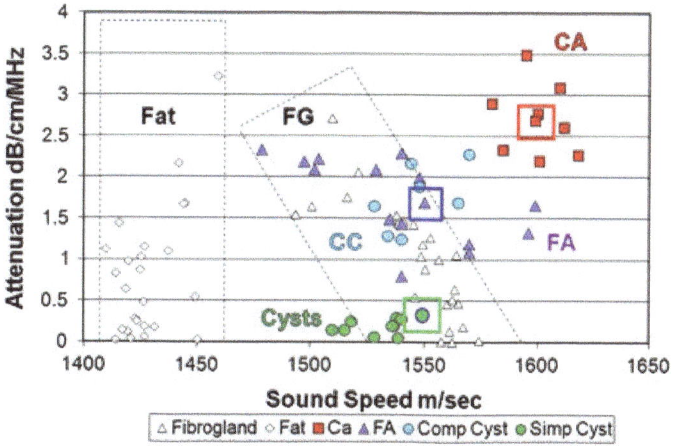

Figure 10: Discrimination of various masses in 125 participants using a commercial prototype system employing full wave inverse scatter reconstructions. Small colored square outlines are for cases shown as examples in [20].

5 Available for Diagnosis

The two primary, commercially available UTT systems are shown in Fig. 11. Both are FDA approved for diagnosis and are in clinical trials for breast cancer screening. No information is expected from those trials until they are completed. Both systems now employ or are beginning to employ full wave inversion reconstructions.

Recent advanced capabilities in commercially available, or soon to be available options include Delphinus's stiffness imaging in multimode displays (Fig. 12) and QT Ultrasound's 3D full wave inversions in times at or approaching clinical utility (Fig 13) [32]. Examples suggest that the stiffness does provide some additional information [33].

With a 1.75D array receiver, QT Ultrasound achieved, in 2017, 3D reconstructions of an entire breast (Fig. 13) in ~35 min, with a scan time of 7 min [32]. When the image plane is at steep angle of incidence to the skin, when near the end of the breast and for relatively flat breasts, diffraction causes large artifacts near the edges of the breast. The outline of the

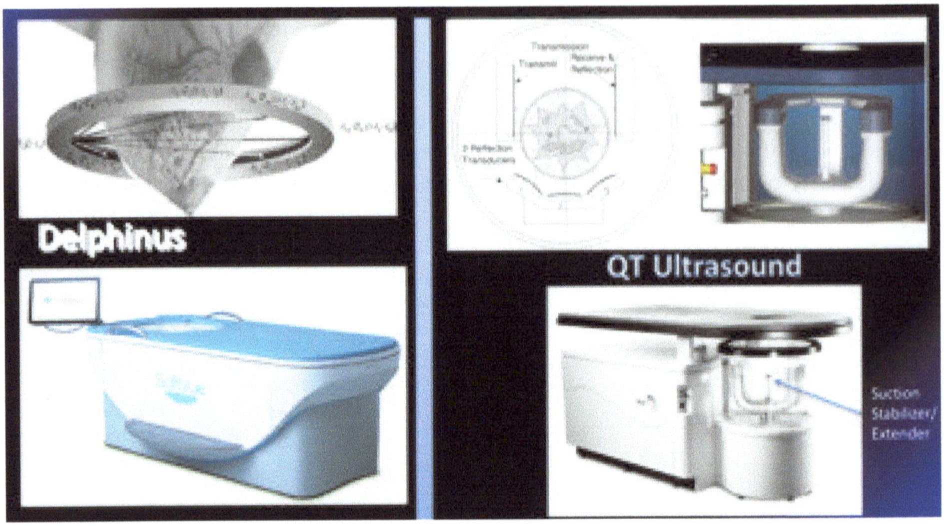

Figure 11: Two commercially available UTT systems. Left: Delphinus SoftVue [31] system with 2048 element ring array operating at up to 2.5 MHz; Right: QT Ultrasound's system with a large, planar transmitter, a 1536 element transmitter and receiver array operating at 0.3 to 1.3 MHz and three conventional curved linear arrays conventional pulse echo imaging [32].

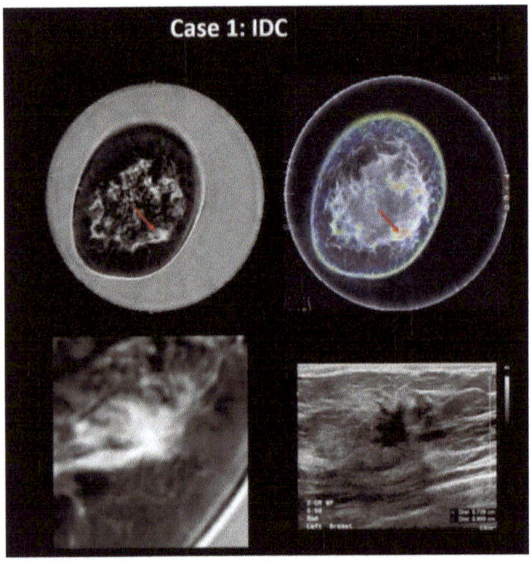

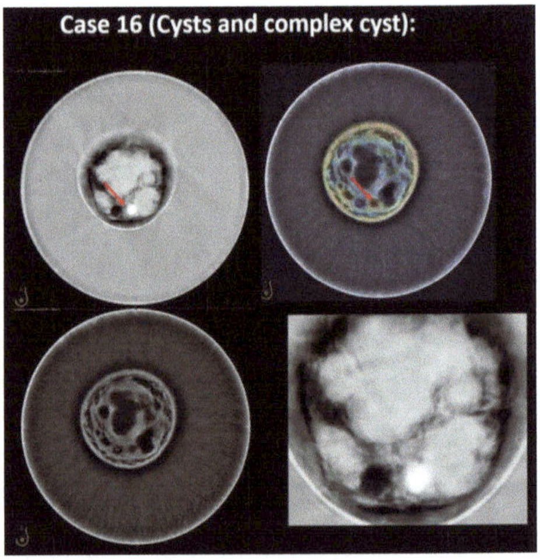

Figure 12: New "stiffness" imaging mode on the Delphinus system. In both cases, SOS and composit reflection mode and SOS-corrected stiffness and attenuation are shown on the top left and right, respectively. Red arrows point to the mass in each. (a) The infiltrating ductal carcinoma is red, i.e., stiff. Left and right bottom are the magnified SOS image and state of the art pulse echo image with a conventional hand held transducer. (b) The complex cyst is somewhat stiff (arrow) and the cysts are not colored, i.e., not stiff nor attenuating. Left and right bottom images are reflection mode and magnified SOS, respectively. Figure adapted from [33].

breast is shrunk in the 2D reconstructions and information is lost, particularly in the critical subareolar area.

In an axial plane of the 3D-reconstruction (top of the lower pair of images) much greater fidelity and detail is revealed in the skin, subcutaneous fat, areolar tissues and nipple. While very little motion of the breast is allowed for these coherent, 3D, full wave reconstructions, that stability is achieveable with a stabilizer/extender post shown in Fig. 11b. Stabilizing tape is required in some or all cases that produced artifacts at the blue arrows in the top images of Fig. 13. Even with data acquired in individual slices in a ring array, reasonable 3D reconstructions appear possible [34]. The small elevational motion of the array stirs the water and the breast less than a rapid rotation as well the elevational motion. Use of custom-formed, semisolid coupling gels instead of water could be a good stabilizer.

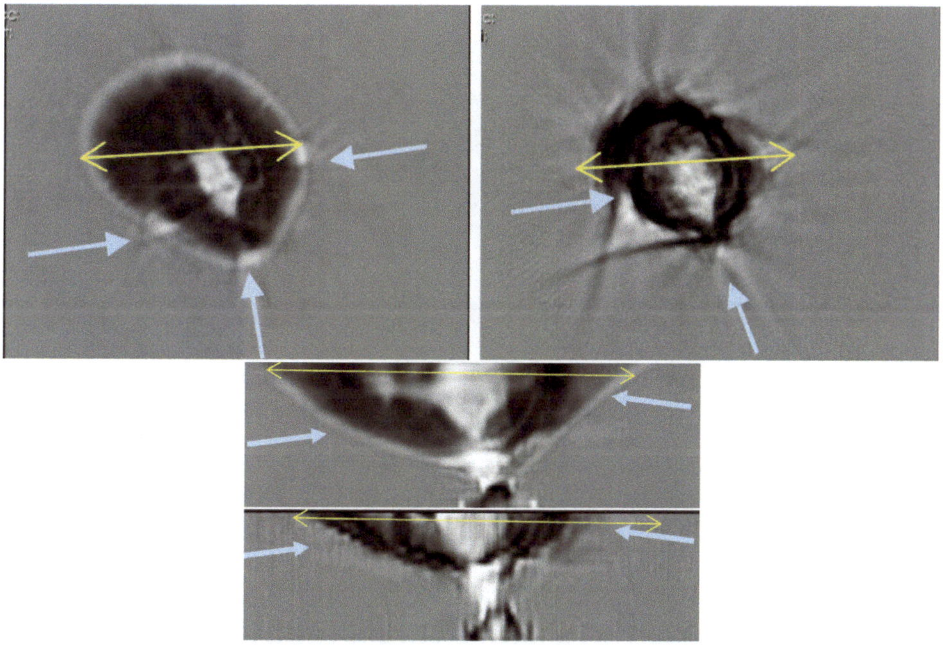

Figure 13: The 90 mm yellow arrows show how, when the image plane is at steep angle of incidence to the skin, the outline of the breast is shrunk in the 2D reconstructions (top right and bottom of the second row) compared with the 3D images (top left and upper image of the bottom row). Reproduced with permission from [32].

6 Other Diverse Systems and Applications

Other diverse systems and applications, too numerous to cover fully here, have been evaluated for breast UTT, many of which are covered elsewhere in these proceedings. Use of a focusing mirror instead of opposed arrays could save some cost [35] and resulted in amazing SOS-corrected reflection mode images in the fatty-breast examples [36]. Both SOS and reflection images were at 7.5 MHz nominal frequency. Phase ambiguity might destroy the SOS quality in dense and heterogeneously dense breasts. It is worth mentioning the near-hemispherical bowl transducer arrays developed for photoacoustic imaging of the breast, superficial tissues and small animals [37,38]. With simple rotation of the bowl about its central axis, and the correct spiral distribution of transducer elements, all of k-space can be sampled with a limited number of elements. This geometry may be equally optimal for UTT.

With high signal to noise ratio and dynamic range and full wave migration, it is becoming possible with UTT to image body parts containing bone. Excellent examples have been obtained with commercial systems in planes containing modestly calcified bone [39] and the adult brain is not far away.

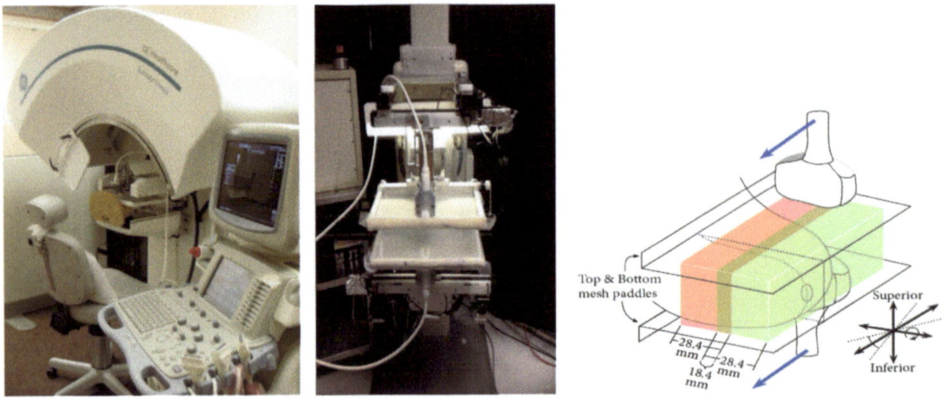

Figure 14: Ultrasound in the mammographic geometry. (a) Combined DBT and pulse echo ultrasound system; (b) Standalone dual sided test system for future inclusion with DBT. Photoacoustics was included in this system. (c) Scanning pattern with one or opposed matrix or linear arrays.

7 Limited Angle Ultrasound Tomography

Our work with GE Global Research, Niskayuna, NY, began with pulse echo imaging of the breast in the mammographic geometry for completely coregistered ultrasound and mammograpy or DBT. It was found that good coverage of the entire breast at frequencies even below 9 MHz could not be achieved by one sided imaging with conventional transducers. Dual sided pulse echo and transmission imaging were evaluated for eventual use with 2D CMUT arrays covering both sides of the entire breast. The CMUT arrays could not be produced with a high enough yield, but the principles were demonstrated with conventional 1.25D arrays (GE M12L). 90% coverage was achieved and better ways of acoustic coupling envisioned to achieve 98-100% coverage. That is better than typically achievable with the breast dependent in water.

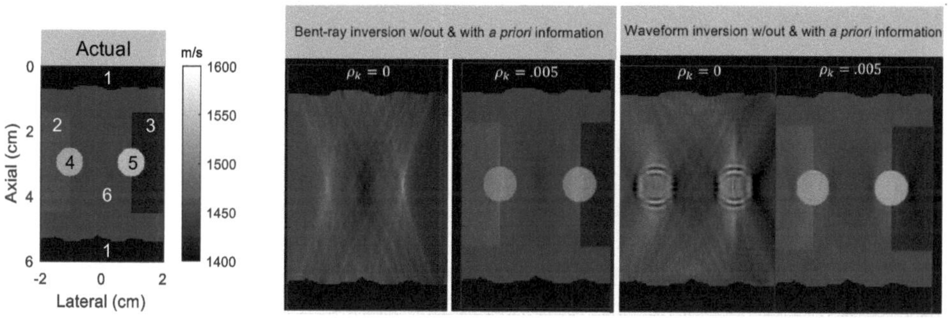

Figure 15: Limited angle transmission tomography of a challenging, compressed breast phantom, shown quantitatively on the left. Bent-ray reconstruction with and without use of a priori information obtainable from reflection mode images [40] are shown on the right, followed by full wave inversion reconstructions with and without use of the a priori information.

For transmission tomography, as well as pulse echo, in the mammographic geometry, the maximum breast thickness with modest compression is 9 cm, compared with a maximum path length of >20 cm for the prone water path. The shorter path length allows higher frequencies for penetration and some reduction in phase ambiguities due to less long adjacent paths with different speeds of sound. With long, opposed linear arrays moving away from the chest wall, as in the single transducer case of [41], good 2D transmission tomography image slices should be achievable *in vivo*, but breast stability for 3D full wave inversion will be difficult unless solid compression paddles can be made to work

Reasonable limited angle reconstructions require use of *a priori* information [42], and such use will be helpful in imaging through and around bone [43].

8 Discussion

Speed of sound tracks rather closely with CT density, so it will be interesting to see whether breast CT can fulfill the role of SOS. Breast CT by itself will not have the added information provided by ultrasound in addition to SOS. That is, μ, physical density, quasistatic and shear wave complex elasticity, reflection and vascularity imaging, relatively safe and simple contrast agent imaging, slip boundary imaging under manipulation and thermoacoustic quantitative imaging. Only with transmission imaging can you have all those ultrasound modes with high spatial and quantitative resolution as well as subwavelength aberration correction of those ultrasound and multimodality imaging modes.

References

[1] Greenleaf, J. F., S. A. Johnson, S. L. Lee, G. T. Herman and E. H. Wood.: "Algebraic reconstruction of spatial distributions of acoustic absorption within tissue from their two-dimensional acoustic projections. in Acoust Hologr and Imaging, Vol 5, Procs Int Symp, Jul, 1973, (1974) 591-603.

[2] Gordon, R., R. Bender, and G. T. Herman (1970): Algebraic reconstruction techniques (ART) for three-dimensional electron microscopy and x-ray photography. J Theor Biol 29:471-481.

[3] Greenleaf, J. F., Johnson, S. A., Samayoa, W. F. and Duck, F. A. (1975) Algebraic reconstruction of spatial distribution of acoustic velocities in tissue from their time-of-flight profiles. In Acoustical Holography, Vol. 6 (Ed. N. Booth), Proc Int Symp Feb, 1975, 71-90. Plenum Press, New York.

[4] Greenleaf, J. F. and S. A. Johnson (1976). "Algebraic reconstruction of spatial distributions of refractive index and attenuation in tissues from time-of-flight and am-plitude profiles." Proc 1st Semin on Ultrason Tissue Charact, May, 1975, (NBS Special Pub. 453): 109-119.

[5] Carson, P. L., T. V. Oughton, W. R. Hendee and A. S. Ahuja (1977). "Imaging soft tissue through bone with ultrasound transmission tomography by reconstruction." Medical Physics 4: 302-309.

[6] Marr, R. B., 1974: On the reconstruction of a function on a circular domain from a sampling of its line integrals. J. of Math. Anal. and Appl., 45, 357-374.

[7] Crawford, C. R. and A. C. Kak (1982). "Multipath artifact corrections in ultrasonic transmission tomography." Ultrason Imaging 4: 234-266.

[8] Greenleaf, J. F. "Effects Of Diffraction And Refraction On Computer-Assisted Tomography With Ultrasound," Proc. SPIE 0372, Physics and Engineering in Medical Imaging, (Nov. 1982); doi: 10.1117/12.934521

[9] Dines, K. A. and A. C. Kak (1979). "Ultrasonic attenuation tomography of soft tissues." Ultrason Imaging 1(1): 16-33.

[10] Azimi, M. and A. C. Kak (1983). "Distortion in diffraction tomography caused by multiple scattering." IEEE Trans Med Imaging 2(4): 176-195.

[11] Twersky, V. (1987). "Low-frequency scattering by correlated distributions of particles." J Acoust Soc Am 81: 1609-1618.

[12] Andersen, A. H. and A. C. Kak (1984). "Simultaneous algebraic reconstruction technique (SART): a superior implementation of the art algorithm." Ultrason Imaging 6(1): 81-94.

[13] A. J. Devaney: A filtered backpropagation algorithm for diffraction tomography. Ultrason Imaging 4 (1982) 336-350.

[14] A. J. Devaney, G. Beylkin: Diffraction tomography using arbitrary transmitter and receiver surfaces. Ultrason. Imaging 6 (1984) 181-193.

[15] Z. Q. Lu, M. Kaveh, R. K. Mueller: Diffraction tomography using beam waves: Z-average reconstruction. Ultrason. Imaging 6 (1984) 95-102.

[16] Duchêne, D. Lesselier, W. Tabbara: Diffraction tomography approach to acoustical imaging and media characterization. Journal of the Optical Society of America A: Optics and Image Science, and Vision 2 (1985) 1943-1953.
[17] Klepper, J. R., G. H. Brandenburger, J. W. Mimbs, B. E. Sobel and J. G. Miller (1981). Application of Phase-Insensitive Detection and Frequency-Dependent Measurements to Computed Ultrasonic Attenuation Tomography. IEEE Transactions on Biomedical Engineering BME-28: 186-201.
[18] Chenevert, T. L., D. I. Bylski, P. L. Carson, C. R. Meyer, R. M. Schmitt, P. H. Bland and D. Adler (1984). "Ultrasonic computed tomography of the breast. Improvement of image quality by use of cross-correlation time-of-flight and phase-insensitive attenuation measurements." Radiology 152: 155-159.
[19] Zeqiri, B., C. Baker, G. Alosa, P. N. T. Wells and H. D. Liang (2013). "Quantitative ultrasonic computed tomography using phase-insensitive pyroelectric detectors." Phys Med Biol 58: 5237-5268.
[20] André, M., J. Wiskin and D. Borup (2013). Clinical Results with Ultrasound Computed Tomography of the Breast. Quantitative ultrasound in soft tissues. J. Mamou and M. L. Oelze. Dordrecht, Springer: 395-432.
[21] R. K. Mueller, M. Kaveh, G. Wade: Reconstructive Tomography and Applications to Ultrasonics. Proceedings of the IEEE 67 (1979) 567-587.
[22] R. J. Lavarello, M. L. Oelze: Tomographic reconstruction of three-dimensional volumes using the distorted born iterative method. IEEE Trans. Med. Imaging 28 (2009) 1643-1653.
[23] R. J. Lavarello, A. J. Hesford, in Quantitative ultrasound in soft tissues, J. Mamou, M. Oelze, Eds. (Springer, Dordrecht, 2013), chap. 14, pp. 345-394.
[24] G. H. Glover: Computerized time-of-flight ultrasonic tomography for breast examination. Ult Med Biol 3 (1977) 117-127.
[25] J. F. Greenleaf, R. C. Bahn: Clinical Imaging with Transmissive Ultrasonic Computerized Tomography. IEEE Trans. Biomed. Eng. BME-28 (1981) 177-185.
[26] J. S. Schreiman, J. J. Gisvold, J. F. Greenleaf, R. C. Bahn: Ultrasound transmission computed tomography of the breast. Radiology 150 (1984) 523-530.
[27] A. L. Scherzinger et al.: Assessment of ultrasonic computed tomography in symptomatic breast patients by discriminant analysis. Ultrasound Med. Biol. 15 (1989) 21-28.
[28] M. P. André, C. H. Barker, N. Sekhon, J. Wiskin, D. Borup, and K. Callahan, "Pre-Clinical experience with Full-Wave Inverse-Scattering for breast imaging acoustical imaging," ser. Acoustical Imaging, I. Akiyama, Ed. Dordrecht: Springer Netherlands, 2009, 29, ch. 10, 73-80.
[29] W. A. Berg et al.: Detection of breast cancer with addition of annual screening ultrasound or a single screening MRI to mammography in women with elevated breast cancer risk. JAMA - J Am Med Assn 307 (2012) 1394-1404.
[30] N. Duric, P. Littrup, in Breast Imaging, C. M. Kuzmiak, Ed. (Intechopin, 2018), chap. 6, pp. 111-139.
[31] Li, N. Duric, P. Littrup, L. Huang: In vivo breast sound-speed imaging with ultrasound tomography. Ultrasound Med. Biol. 35 (2009) 1615-1628.

[32] J. W. Wiskin, D. T. Borup, E. Iuanow, J. Klock, M. W. Lenox: 3D Nonlinear Acoustic Inverse Scattering: Algorithm and quantitative results. IEEE Trans. Ultrason. Ferroelectr. Freq. Control (2017) 1161-1174.

[33] Li, G. Sandhu, M. Boone, N. Duric, P. Littrup and K. Bergman: Breast tissue characterization with sound speed and tissue stiffness imaging. in Medical Imaging 2018. (2018), Proc. SPIE 10580, pp. 105800Y.

[34] G. Y. Sandhu, E. West, C. Li, O. Roy, N. Duric, in Medical Imaging 2017: Ultrasonic Imaging and Tomography. (2017), vol. 10139.

[35] J. Nebeker, T. R. Nelson: Imaging of sound speed using reflection ultrasound tomography. J. Ultrasound Med. 31 (2012) 1389-1404.

[36] M. Hollenhorst, C. Hansen, N. Hüttebräuker, A. Schasse, L. Heuser, H. Ermert, G. Schulte-Altedorneburg: Ultrasound Computed Tomography in Breast Imaging: First Clinical Results of a Custom-Made Scanner. Ultraschall in Med 31 (2010) 604-609.

[37] R. A. Kruger, R. B. Lam, D. R. Reinecke, S. P. Del Rio, R. P. Doyle: Photoacoustic angiography of the breast. Med. Phys. 37 (2010) 6096-6100.

[38] R. A. Kruger et al.: Dedicated 3D photoacoustic breast imaging. Med. Phys. 40 (2013).

[39] J. Wiskin, e. al, in 2019 IEEE Int Ultrasonics Symp. (IEEE, Glasgow, 2019), pp. 951-958.

[40] D. Larson, W. M. Lee, M. A. Roubidoux, M. M. Goodsitt, C. Lashbrook, C. E. Davis, O. D. Kripfgans, P. L. Carson: Preliminary Clinical Experience with a Combined Automated Breast Ultrasound and Digital Breast Tomosynthesis System. Ult Med Biol 44 (2018) 734–742.

[41] M. Hooi, P. Carson: First-arrival Traveltime Sound Speed Inversion with A Priori Information. Med. Phys. 41 (2014) 548-561.

[42] B. C. Byram, N. Duric, S. Shiguemi Furuie, D. A. Cardona Cardenas, paper presented at the Medical Imaging 2018: Ultrasonic Imaging and Tomography, 2018.

[43] R. Jintamethasawat, Univ. of Mich., Dissertation, Ann Arbor (2018) Ch 7, 150-169.

Ultrasound Imaging with FWI: From Breast to Brain

Lluís Guasch, Ph.D.[1]

[1] *Imperial College London, Department of Earth Science and Engineering, London, United Kingdom, E-mail: l.guasch08@imperial.ac.uk*

Abstract

Full-waveform inversion (FWI) is an imaging technique developed in the field of seismology that exploits all available information in the data, phase and amplitude, by solving a local optimisation problem based on the numerical solution of the wave equation. This technology was first translated to medical breast imaging over a decade ago. It has dramatically improved the potential of ultrasound as an imaging tool for breast cancer diagnosis due to its ability to produce high-resolution images and to provide quantitative information of several tissue properties such as acoustic speed, density, impedance or attenuation.

Despite its many challenges, breast imaging with FWI benefits from the low tissue heterogeneity of the target and from adequate instrument access to provide sufficient illumination. Brain imaging, on the other hand, presents a more challenging problem due to the presence of the skull. The high-contrast bone tissue surrounding the imaging target, i.e. the brain, requires full 3D data acquisition as well as some a priori knowledge of the geometry and acoustic properties of the skull. Under the right circumstances, however, brain imaging with FWI is possible and has the potential to impact the diagnosis and monitoring of a wide range of neuropathologies like stroke, brain cancer or head trauma.

Biography
Dr Lluís Guasch is a Research Fellow at Imperial College London. He performs his research across the Earth Science and Engineering, and the Bioengineering departments. He holds a B.Sc. in physics from the University of Barcelona and a Ph.D. in Geophysics from Imperial College London. His work in seismic imaging has let to a successful start-up company, S-Cube, where he is one of the founders and the COO, as well as author of the two patents that drive the technology and commercial development of the company. During the last four years his research has been focused on medical imaging, and currently leads an academic multi-disciplinary group that works on translating seismic imaging technologies to ultrasound imaging for clinical applications.

Full-ring Photoacoustic Tomography: Light and Sound to Enhance Diagnosis of Breast Cancer

Mohammad Mehrmohammadi, Ph.D.[1,2,3]

[1] *Wayne State University, Department of Biomedical Engineering, Detroit, USA*
E-mail: mehr@wayne.edu
[2] *Wayne State University, Department of Electrical and Computer Engineering, Detroit, USA*
[3] *Barbara Ann Karmanos Cancer Institute, Detroit, USA*

Abstract

Photoacoustic imaging has shown a steadily growth in diagnostic imaging of various pathologies including cancer. Through excitation of the tissue with light, conversion of the light energy to thermal energy, minor but rapid thermoelastic expansion followed by generation of acoustic waves, PA provides a complementary platform to acquire optical signature of breast tumors using the same hardware used for US imaging. In recent years, PA tomography (PAT) imaging of breast cancer has shown a steady growth. We have developed a PAT system based on a ring geometry (both excitation and detection) that can potentially address limitations of existing PAT system. Within this presentation, an overview of PAT application in breast cancer detection and staging as well as initial results from our developed full-ring PAT system will be presented.

Systems

The New Generation of the Breast Ultrasound Tomography Imaging System in HUST

J. Song[1], K. Liu[1], Q. Zhang[1], L. Zhou[1], Q. Zhou[1], Z. Quan[1], Y. Wu[1], X. Fang[1], M. Yuchi[1], and M. Ding[1]

[1] *Huazhong University of Science and Technology, Wuhan, China*
E-mail: junjiesong91@hust.edu.cn

Abstract

Background
In 2017, our team developed the prototype of the ultrasound tomography system. To this end, we have designed and implemented a new generation of the ultrasound tomography system to meet the clinical requirement.

Material and Method
The new system mainly includes a patient bed, image reconstruction servers and a display workstation. A ring array probe is adopted, which have 2048 elements with the center frequency 3.0 MHz and the radius 110 mm. The system has 512 independent transmit and receive channels. The excitation voltage can be up to +/-100V. After digitization, the RF signals are transmitted to the reconstruction servers, and the GPU completes the image reconstruction. The display station receives and displays the reconstructed images and provides an interface for data management and debugging.

Results
The acquisition time for one slice can be set between 2 s to 8 s depending on different imaging modes. The reconstruction time of the reflection image of 2048x2048 pixels is 4s, and the reconstruction time of the sound speed image and the attenuation image of 1024x1024 pixels are 20s and 25s.

Conclusion
Preliminary tests have shown that the performance of the new system has been significantly improved. The future work is to optimize the system based on clinical results.

Progress Towards an Open-Source, Low-Cost Ultrasound Computed Tomography Research System

Morgan J. Roberts[1], Eleanor Martin[1], Ben T. Cox[1], and Bradley E. Treeby[1]

[1] *Department of Medical Physics and Biomedical Engineering, University College London, London, UK, E-mail: morgan.roberts.18@ucl.ac.uk*

Abstract

Ultrasound Computed Tomography (UCT) systems are typically custom-built, and the high cost and long lead time can be a significant barrier to research groups becoming active in the field. This work details progress towards FlexUCT: a design framework for a low-cost benchtop UCT system that will allow novel data acquisition protocols and reconstruction algorithms to be tested on a physical system, thereby facilitating progress towards fast and accurate UCT. The design, assembly and testing of 4-element prototypes are described. Field scans of the prototypes have been performed to characterise the beamwidth, element radiation pattern and acoustic cross talk, and the pulse echo behaviour was evaluated.

Keywords: Ultrasound Computed Tomography, Low Cost Hardware, Open Source, US Transducer

1 Introduction

Ultrasound Computed Tomography (UCT) is a biomedical imaging technique that involves sending pulses of ultrasound into the tissue, often the breast, from different angles, and measuring the reflected and transmitted pulses with an array of detectors [1]. UCT has the advantages of being non-ionising, operator-independent, and pain-free since it requires no breast compression [2].

However, for researchers seeking to start working on UCT the barrier to entry is high, because to properly test novel data acquisition protocols and image reconstruction methods, custom hardware is required in order to have sufficient control over the experimental setup to meet the specific research needs, and custom systems can be prohibitively expensive and often have a long lead time. Lowering this barrier will increase participation in the field, which is likely to result in more and faster progress towards improved image reconstruction algorithms and data acquisition protocols.

This paper describes the initial developmental stages of FlexUCT, an open source design framework for a benchtop UCT research system. The framework uses a toolbox of flexible, low-cost manufacturing techniques which allow the user to choose their own values for key design parameters, such as aperture size and element dimensions, without affecting the assembly processes. The framework is designed to have simple user requirements for manufacture, requiring only standard workshop hand tools, a mill and a 3-D printer, which are commonly available in most research institutions. Our future goal is to distribute the complete specification needed to build a custom UCT research system, including parametric CAD models, 3D-printing design files, Gerber files for PCB manufacture, a bill of materials and a full assembly procedure. Here, we present the design, assembly, and evaluation of the first FlexUCT prototypes.

2 Design and Modelling

2.1 Aperture Configuration

To develop flexible manufacturing techniques, our approach was to first design a typical full-aperture UCT ring, but to build this in smaller 4-element segments. This allows rapid iteration with lower material costs, and also ensures that the manufacturing techniques can be used to realise modular apertures as well as fixed apertures. The KIT and MUBI UCT systems demonstrate that multiple modular segments can be assembled to realise 2D [3] and 3D [4] apertures, which could provide greater flexibility to FlexUCT users by allowing changes to the aperture configuration without manufacturing new hardware, and by allowing easy replacement of faulty elements.

The CURE prototype [5] has previously shown that 256 transducers are sufficient for basic UCT imaging in a 2D ring configuration, and since 3D apertures require an order of magnitude more transducers, meaning complicated and expensive multiplexing would have to be used, it was decided to use a 2D ring array configuration with 256 equispaced transducers as shown in Figure 1. This matches the number of available transmit/receive channels for the ULA-OP, UARP, and Verasonics Vantage systems, which are three commonly available platforms for ultrasound data acquisition [6]. In principle the system could be rotated to obtain measurements at a finer angular spacing. A ring diameter of 220mm was chosen, based on an assessment of breast diameter from 200 women in America showing that a ring of this size would accommodate 95% of the population [7], which is shown in Figure 1 with the PZT element dimensions.

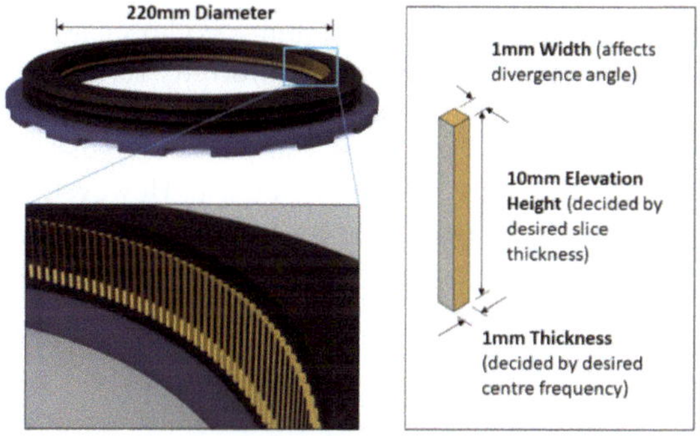

Figure 1: Left: Full-aperture UCT ring, designed with 256 equispaced transducer elements embedded in a common backing layer, and with a diameter of 220 mm. Right: A single PZT element defined by the key dimensions shown, with solid pattern electrodes.

2.2 PZT Element Dimensions

The centre frequency of a PZT element is dominated by its half-wave resonance, which is controlled by its thickness when driven as a width expander [8], and should be as high as possible for good resolution, but low enough to prevent excessive absorption. For a plane wave travelling in the $+z$ direction, the solution to the lossy Helmholtz equation is $p = P_0 e^{-\alpha z} e^{j(\omega t - kz)}$ [9], where p is the acoustic pressure, ω is the angular frequency, k is the wavenumber, and t is time. This shows that as a wave propagates through an absorbing medium over a distance z, the plane wave amplitude P_0 decays according to $\exp(-\alpha z)$. In tissue, the frequency dependent acoustic absorption coefficient $\alpha(f)$ follows a power law relationship described by $\alpha(f) = \alpha_0 |f|^y$, where $\alpha_0 = 0.75$ dB.MHz^{-y}.cm^{-1} and $y = 1.5$ for breast tissue [8]. Using these material properties, a mean breast diameter of $z = 140$ mm, and an assumed noise floor of 5%, a centre frequency of $f = 2$ MHz results in an acceptable decay, which corresponds to a PZT element thickness of 1 mm with a sound speed of 4010 m.s^{-1} [10].

The elevation height of an element affects the beamwidth and therefore the image slice thickness. To avoid variations in the effect of volume averaging, the slice thickness would ideally be constant over the imaged region [2]. To choose an appropriate elevation height, simulations were performed using the acoustic field propagator from the k-Wave toolbox in MATLAB [11] [12]. For a range of element sizes, the transmitted field in the elevation plane was computed at a frequency of 2 MHz for a ring of diameter 220 mm. As the receive sensitivity will have the same spatial dependence, the transmit-receive sensitivity was found by multiplying the transmitted field with a copy of itself flipped about the central point of the array. The resulting transmit-receive sensitivity is shown in Figure 2. The full-width-half-maximum of the beam at the centre of the array was then extracted. As the element elevation height increases, the beamwidth at the centre decreases as expected, but beyond the 10 mm element size, the improvement in beamwidth comes at the expense of a much wider and longer near field which could affect the image reconstruction at the edges of phantoms larger than 100 mm. An element elevation height of 10 mm was chosen as a good balance between slice thickness at the centre of the array and slice uniformity across the region of interest.

For UCT, the elements would ideally be as narrow as possible so that they are very diverging in the scan plane and achieve full coverage of the breast. However, it is important for this system to be easily and repeatably assembled by hand, so the element width was chosen to be 1 mm, so that the electrode area is large enough to easily terminate signal leads by hand-soldering. The final element dimensions can be seen in Figure 1.

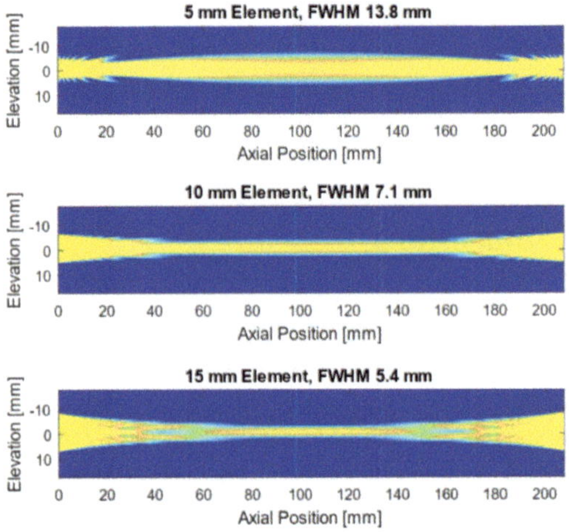

Figure 2: Elevation plane transmit-receive sensitivity plots for rectangular bar elements driven at 2 MHz, for a range of elevation heights. Pressure is displayed on a logarithmic scale, referenced to the maximum pressure at the centre of the ring, and thresholded to a 6 dB range. The element width was 1 mm for all simulations.

2.3 Acoustic Properties of Tungsten-Epoxy Composites

To fabricate matching and backing layers, tungsten particles can be embedded in an epoxy resin polymer matrix to form a 0-3 composite. A study was performed to establish the relationship between tungsten-epoxy weight ratio and the acoustic properties of the resulting composite under a specific set of repeatable conditions, since previous work has only been completed at much higher frequencies [13], and work by Grewe et al. suggests that the relationship is affected by tungsten particle size and mixing techniques [14].

Tungsten-epoxy samples were manufactured with tungsten weight ratios ranging from 72.5% to 95%. To calculate the characteristic impedance of the samples, their sound speed was measured in deionised water using the through-transmission substitution method [15], and their density was calculated using a simplified hydrostatic weighing method [16]. Figure 3 shows the measured sound speed and characteristic impedance, alongside the predicted acoustic properties of the samples calculated using the Devaney model, which gives expressions for the effective bulk and shear moduli of a random distribution of elastic spheres embedded in an elastic matrix, using multiple scattering theory [17]. The shape of the predicted sound speed profile is qualitatively similar to the experimental data, showing a sound speed minimum at a tungsten weight ratio of 87.5%, but the model does not match the data quantitatively since it doesn't account for absorption, and since it also depends heavily on the values used for the bulk and shear moduli of the epoxy matrix and tungsten particles. The experimental results are important because they can be interpolated to manufacture tungsten-epoxy matching and backing layers with a desired characteristic impedance.

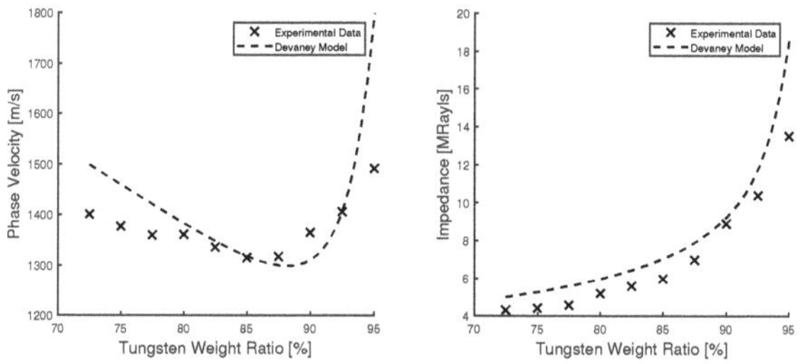

Figure 3: Experimental data showing the relationship between the tungsten weight ratio of a tungsten-epoxy composite and its phase velocity and characteristic impedance at 2 MHz. The acoustic properties from the Devaney model [17] are also shown.

2.4 CAD Design

The prototype transducers were designed in CAD software, shown in Figure 4. PZT elements with a single tungsten-epoxy matching layer on their front face are embedded in a 12 mm tungsten-epoxy backing layer, which is mounted on an acrylic cable-guide plate, which has holes to align the individual signal leads with the rear electrode on each element. The individual signal leads combine into a flat ribbon cable, which is secured to the cable-guide plate with a strain relief clip. A common ground electrode connects to a 1.5×1 mm unmatched area on the front electrode of each PZT element.

3 Assembly Methods

3.1 Quarter-Wave Matching Layer

For a layer of tungsten-epoxy used to match PZT with impedance $Z_p = 30.5$ MRayl to water with impedance $Z_w = 1.5$ MRayl, the impedance of the matching layer should be $Z_l = \sqrt{30.5 \times 1.5} = 6.76$ MRayl. This can be achieved by using a tungsten weight ratio of $W_T = 86.9\%$, which has a sound speed of 1317 m.s^{-1}, meaning that the target quarter wave matching layer thickness is $\lambda/4 = 165\ \mu m$ at 2 MHz.

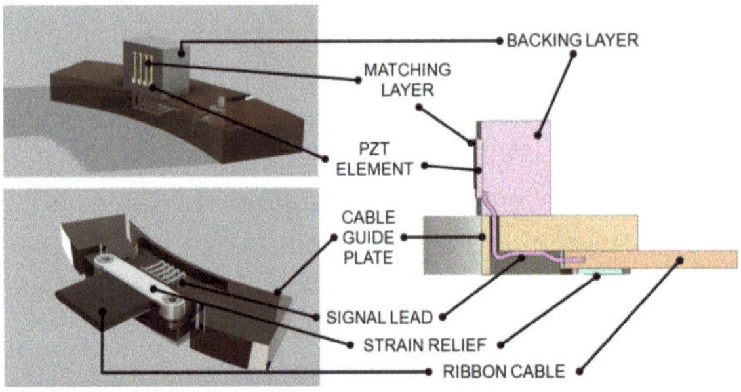

Figure 4: CAD rendering of prototypes, with the key components labelled. The right hand diagram is an elevation plane cross section down the centre of one of the PZT elements. Note that the common ground electrode on the front face of the PZT elements is not shown.

Previous low cost transducers have used a 3D-printed mould to control the matching layer thickness by supporting the PZT element on 'standoffs' which are exactly $\lambda/4$ [18]. Preliminary work showed that as frequency increases, matching layer thicknesses becomes less repeatable, because the required thickness approaches the layer height of the 3-D printed standoffs, meaning that the technique is not suitably flexible. Alternatively, it has been shown that metal foils can be used as a stencil to deposit solder pastes with a uniform and controlled geometry for circuit board assembly [19], and that blade coating can be used for depositing thin films with a controllable thickness [20].

In order to replicate a hybrid of these techniques, a stencil was manufactured by using a 1 mm end mill to create 1×8.5 mm slots in 0.2 mm steel shim stock. A jig was milled from acrylic with 1 mm slots to receive the PZT elements and hold their top face flush with the surface. The stencil was placed on top of the elements using registration pins, and the tungsten-epoxy composite was spread over the slots using a blade. On removal of the stencil, the tungsten-epoxy composite covered the PZT element with a controlled geometry, and was left to cure at room temperature. This process is shown in Figure 5.

To test the repeatability of the technique, 45 samples were deposited and the thickness was measured with digital callipers. The mean sample thickness was 150.9 μm, and the standard deviation was 15.4 μm, placing the 165 μm target thickness within one standard deviation of the mean. Future work will be done to establish the relationship between the stencil thickness and the final cured thickness of the deposited layers, but this preliminary study suggests that the proposed method can deposit matching layers repeatably with a suitable thickness.

3.2 PZT Element Alignment Mould

It is important that the front faces of the elements are aligned tangent to the ring geometry. If they are skewed, then the breast may not be homogeneously insonified, which can cause artefacts [21]. Moulds with slots to retain each element were manufactured from polyvinyl

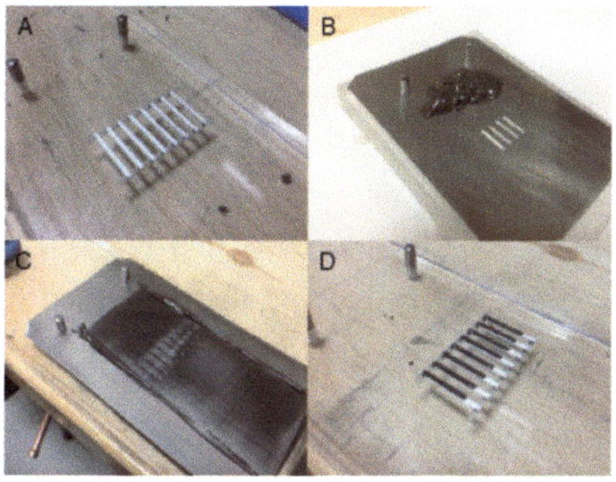

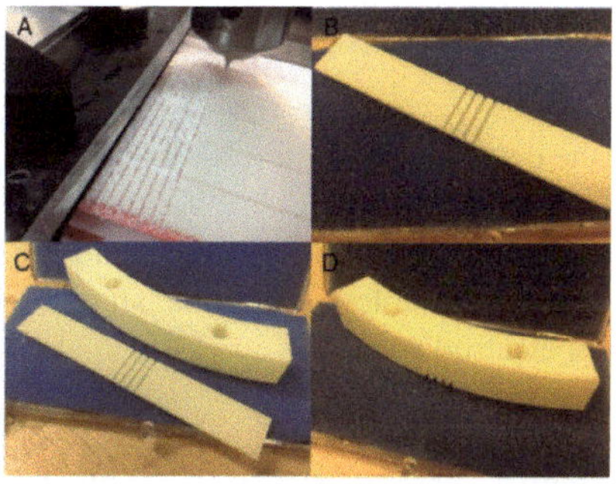

Figure 5: Left: Proposed method used to deposit matching layers. A) PZT Elements are placed into slots in a jig, with their top face flush with the surface. B) A steel stencil is placed on top, and the tungsten-epoxy composite is added. C) The tungsten-epoxy composite is spread over the stencil with a scraper. D) The stencil is removed. Right: Proposed method to manufacture an alignment mould. A) $1 \times 1 \times 13$ mm slots are milled into a strip of 1.5 mm PVA. B) The resulting PVA alignment strip. C) Strip and 3D-printed alignment form before bending. D) The PVA alignment strip is bent and welded onto the alignment form.

alcohol (PVA) in order to align the PZT elements relative to one another. PVA is a water soluble polymer and can easily be removed without damaging the elements by dissolving the moulds. Kerfs were milled from a 3D-printed flat PVA strip, which was then bent and welded to a 3D printed PVA form, shown in Figure 5. The benefit of using milled slots is that the internal corners are much sharper than those achieved using 3D-printing alone, allowing a much tighter fit to be achieved. To test the PZT element orientation accuracy when using milled slots rather than 3D-printed slots, both methods were used to prepare a 4-element sample, and a microCT scan was performed on each using a XT H 225 scanner (Nikon). A slice was taken through the CT image to show the cross sections of the PZT elements, which can be seen compared to microscope photographs of the samples in Figure 6. This shows that the element alignment achieved using milled slots is more uniform than that achieved using 3D-printing alone.

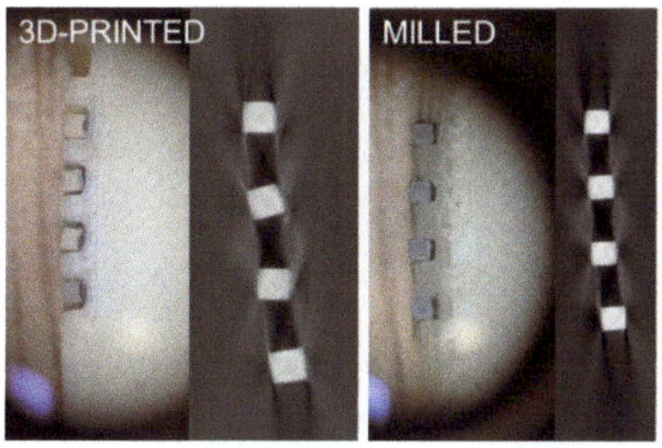

Figure 6: Analysis of the PZT element orientation using microscope photographs (Left) and microCT scans (Right). Two samples were investigated that use different assembly methods to create slots that the PZT elements are placed in. The 3D-printed slots are shown on the left, and the milled slots are shown on the right.

3.3 Further Assembly

Once the deposited matching layers had cured, the PZT elements were fixed in the alignment moulds using a water soluble PVA glue. The flat ribbon cable was then fixed into place with a 3D-printed PLA strain relief clip, and the individual signal leads were distributed through the holes in the cable guide plate, shown in Figure 7. The leads were then soldered to the element electrodes.

A mould for curing the backing layer was 3D-printed from PVA filament. The backing layer mould, cable guide plate, and alignment mould were then all bolted down to a rigid base plate, shown in Figure 7. A tungsten-epoxy composite with a tungsten weight ratio of 92.5% was mixed thoroughly, pushed into the backing layer mould, and allowed to cure at room temperature, bonding directly onto both the rear electrode of the PZT and the cable guide plate, whilst encapsulating the signal leads. A weight ratio of 92.5% is the highest impedance composition available using hand mixing, since beyond this high pressures and temperatures are required to bind the composite together [22]. This was chosen in order to investigate whether the resulting sensitivity is high enough for use in UCT.

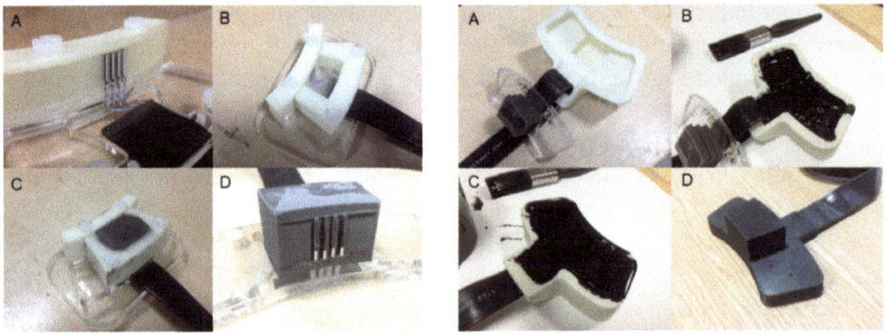

Figure 7: Left: Casting the backing layer. A) Elements are placed in alignment moulds and leads are soldered onto electrodes. B) Backing layer mould fixed into place. C) Tungsten-epoxy backing layer pressed into mould. D) PVA moulds dissolved, leaving elements embedded in backing layer. Right: Coating the transducer. A) Flat ribbon cable threaded through a slot in the coating mould. B) Mould coated with Aptflex F7. C) Transducer placed in the mould and filled up with Aptflex before leaving to cure. D) Finished second generation prototype after PVA mould has been dissolved away.

The PVA moulds were all removed by dissolving them in warm water. In order to waterproof the transducers and to insulate them electrically, a two-component polyurethane coating with an acoustic impedance of 1.5 MRayl was used (Aptflex F7, Precision Acoustics) to minimise the pressure reflection coefficient at the coating-water boundary, preventing internal reverberations within the coating layer. A 3D-printed PVA mould was used to create a coating layer with a consistent thickness, which was again removed by dissolving in water. One of the finished transducers can be seen in Figure 7.

4 Experimental Evaluation

4.1 Field Scan Procedure

Field scans were performed on the prototypes to investigate the pressure distribution on the front of the elements and the beam orientation in the scan plane for each element. Measurements were taken in a tank of deionised water using an X-Y stepper motor positioning system (Precision Acoustics), and the water temperature was monitored using a J-type thermocouple (National Instruments). The transducer was directly connected via a custom 3D-printed backshell to the Vantage-256 (Verasonics), with no electrical impedance matching, and all four channels were driven by a 3 cycle tri-state pulse with a frequency of 2 MHz and an amplitude of 30 V. Signals were acquired at a sampling frequency of 0.2 GHz using a calibrated 200 μm needle hydrophone (Precision Acoustics) connected to an InfiniiVision DSOX3024A digital oscilliscope (Keysight) via a submersible preamplifier and DC coupler (Precision Acoustics) in a 170×170 scanning grid located 27 mm from the transducer surface, with a uniform point spacing of 0.3 mm. The voltage signals were deconvolved using the hydrophone response to give the pressure time series in the measurement plane, which were bandpass filtered. A Tukey window was applied, and the 3D volume of the maximum pressure in the field was calculating by backprojecting using the angular spectrum method [23].

4.2 Elevation Plane Beam Orientation Analysis

To evaluate the orientation of the beams in the elevation plane of the pressure field, the scanning procedure was repeated four times, successively driving each element in isolation. The beam orientation was calculated for each element by drawing a line between the locations of the maximum pressure at two points in the far field, and calculating the angle between this line and the horizontal. This can be seen in Figure 8, which shows the pressure on a logarithmic scale. Although there is a common downwards skew due to misalignment of the scan plane with the transducer, the standard deviation of the beam angles is $\sigma = 0.57°$, which corresponds to a 1.1 mm vertical deviation over a horizontal distance of 110 mm to the centre of the ring. This is sufficiently small to suggest that the milled alignment strip is a suitable low-cost method for aligning the PZT elements relative to one another with high repeatability. The measured beamwidth at the centre of the ring is 9.5 mm, which is slightly greater than the 7.1 mm predicted from the k-Wave simulations, which suggests that the effective element elevation height is smaller than 10 mm.

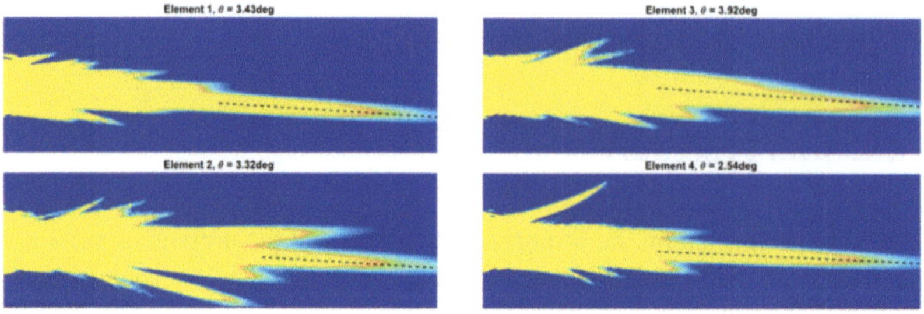

Figure 8: Elevation beam pattern for each of the elements. The centreline of the beam in the far field has been estimated, and the angle of the centreline to the horizontal has been calculated. Pressure is displayed on a logarithmic scale, referenced to a point on the beam axis in the far field, and thresholded to a 6 dB range.

4.3 Element Radiation Pattern and Cross Talk

The front view in Figure 9 shows that each element has a weakly radiating region at the bottom. This is because a 1.5 mm $\times$ 1 mm area was left unmatched to allow a common ground lead to be terminated to the front electrode. This resulted in poor ultrasound transmission from this region, which is likely the reason for the smaller effective elevation height. The width of the radiation pattern varies along the length of the element, which is due to non-uniform coverage caused by a burr on the stencil. Additional lateral non-uniformity in the radiation pattern may be introduced by deformation due to surface tension at the edges of the matching layer during curing. To address these factors in future work, the stencil slots will be enlarged and a slightly thicker matching layer will be deposited and lapped down to size after curing.

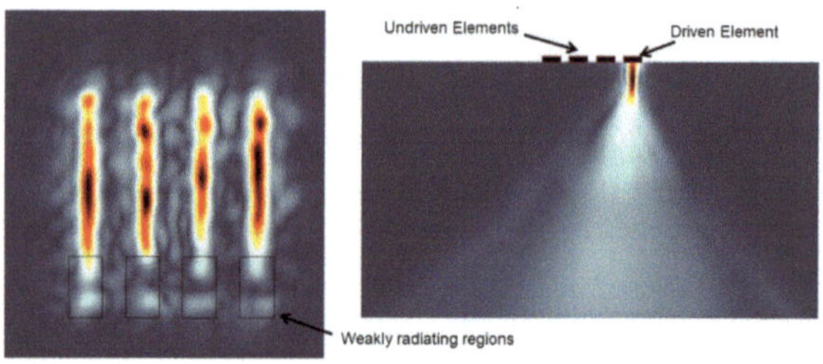

Figure 9: Left: Maximum source pressure showing the radiation pattern of the elements. A weakly radiating region can be seen at the bottom of each element. Right: Maximum pressure in the scan plane of the elements, showing that the undriven elements do not radiate when a single element is driven in isolation.

The pressure fields shown in Figure 9 show that the crosstalk between elements is negligible since none of the non-driven elements are radiating, indicating that there is no electrical interaction between elements, and that there is no acoustic interaction between elements through their common backing layer.

4.4 Pulse Echo Behaviour

To assess whether there were any internal reverberations in the transducers, they were placed in a tank of deionised water facing the wall of the tank, connected to the Vantage-256 (Verasonics), and driven with a 3 cycle tri-state pulse at a frequency of 2 MHz and an amplitude of 30 V. In this setup the tank wall acts as a reflective target. A representative time series is shown in Figure 10, which features a high amplitude burst from electrical pickup, followed by a low amplitude reflection, followed by two high amplitude reflections. The high amplitude reflections are known to arrive from the front and back of the tank wall, because as the distance between the transducer and the target is increased, the arrival times of these pulses increase. The low amplitude pulse was stationary in time when the transducer was moved, indicating that there may be internal reverberations within the transducer. The sound speed of the backing layer was measured to be $c = 1407$ m.s^{-1} and the distance of one round trip within the backing layer is $d = 2 \times 12$ mm $= 0.024$ m. Using this data would result in a pulse arrival time of $\delta t = 0.024/1407 = 17.1$ μs, and the actual pulse arrival time from Figure 10 is 18.5 μs. Allowing for the uncertainty in the backing layer thickness and sound speed, this suggests that the low amplitude reflection is from the backing-Aptflex boundary at the rear face of the backing layer.

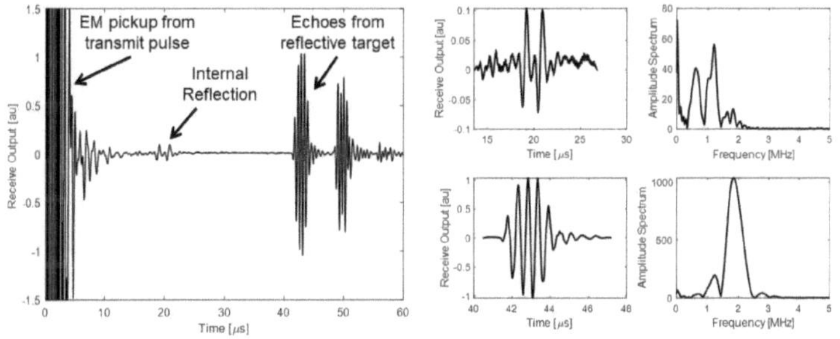

Figure 10: Left: Pulse-echo signal showing electrical pickup up until 10 μs, a low amplitude reflection from the rear face of the backing layer at 18.5 μs, and two high amplitude reflections from the front and rear face of the tank wall at 42 μs. Middle: Extracted time series for the internal reflection (top) and echoes from reflective target (bottom). Right: Amplitude spectrum for the internal reflection (top) and echoes from reflective target (bottom).

The internally reflected pulse and the first pulse to arrive from the external target were both extracted and windowed before applying a Fourier transform. Figure 10 shows that the amplitude spectrum of the internally reflected pulse contains most of its energy up to 1.2 MHz, whereas the amplitude spectrum of the pulse arriving from the external reflective target contains most of its energy at the transducer resonance frequency of 2 MHz. High frequencies are attenuated more strongly when propagating through absorbing media like tungsten-epoxy composites, so it is to be expected that the internally reflected pulse would contain most of its energy at these lower frequencies.

The internal reflections within the prototype transducers are a problem for UCT, since an incoming acoustic pulse will have an initial arrival time when it first reaches the PZT, but will have a second arrival time when it propagates past the PZT, reflects from the rear face of the backing and returns to the PZT. If the second arrival has a high enough amplitude, this could present an artefact during the image reconstruction.

5 Conclusions and Future Work

In this work, the first FlexUCT prototypes have been assembled and evaluated. Novel low-cost methods to deposit matching layers with a repeatable thickness and align PZT elements during manufacture were used to build robust, functioning 4-element transducers, which required no special equipment other than a mill, 3D-printer, and standard workshop equipment. In future, the elevation beamwidth of the transducers will be reduced by increasing their effective elevation height with a better strategy for making electrical connections to the front face of the elements, and internal reverberations in the transducers will be addressed by increasing the absorption of the backing layer. In the future, the release of an open-source design framework for a low-cost benchtop UCT system will enable more research into the optimal configuration for acquiring the minimum amount of data needed for fast, accurate UCT.

6 Acknowledgements

The authors would like to thank Francesco Iacoviello (Chemical Engineering, UCL) for running the microCT scan. This work was supported by the National Physical Laboratory, European Union's Horizon 2020 research and innovation programme H2020 ICT 2016-2017 under Grant Agreement No. 732411 (as an initiative of the Photonics Public Private Partnership), the Engineering and Physical Sciences Research Council (EPSRC), UK, Grant Nos. EP/L020262/1, EP/S026371/1, and EP/P008860/1, and the EPSRC Centre For Doctoral Training in Intelligent Integrated Imaging In Healthcare (i4health).

References

[1] C. Li, N. Duric, and L. Huang, "Clinical breast imaging using sound-speed reconstructions of ultrasound tomography data," in *Medical Imaging 2008: Ultrasonic Imaging and Signal Processing*, vol. 6920, p. 692009, International Society for Optics and Photonics, 2008.

[2] N. Duric, P. Littrup, C. Li, O. Roy, S. Schmidt, R. Janer, X. Cheng, J. Goll, O. Rama, L. Bey-Knight, *et al.*, "Breast ultrasound tomography: bridging the gap to clinical practice," in *Medical Imaging 2012: Ultrasonic Imaging, Tomography, and Therapy*, vol. 8320, p. 83200O, International Society for Optics and Photonics, 2012.

[3] N. V. Ruiter, G. Göbel, L. Berger, M. Zapf, and H. Gemmeke, "Realization of an optimized 3d usct," in *Medical Imaging 2011: Ultrasonic Imaging, Tomography, and Therapy*, vol. 7968, p. 796805, International Society for Optics and Photonics, 2011.

[4] J. Camacho, J. F. Cruza, N. González-Salido, C. Fritsch, M. Pérez-Liva, J. Herraiz, and J. Udías, "A multi-modal ultrasound breast imaging system," in *Proceedings of the International Workshop on Medical Ultrasound Tomography: 1.- 3. Nov. 2017, Speyer, Germany*, pp. 119–130, KIT Scientific Publishing, 2018.

[5] N. Duric, P. Littrup, L. Poulo, A. Babkin, R. Pevzner, E. Holsapple, O. Rama, and C. Glide, "Detection of breast cancer with ultrasound tomography: First results with the computed ultrasound risk evaluation (cure) prototype," *Medical physics*, vol. 34, no. 2, pp. 773–785, 2007.

[6] E. Boni, C. Alfred, S. Freear, J. A. Jensen, and P. Tortoli, "Ultrasound open platforms for next-generation imaging technique development," *IEEE transactions on ultrasonics, ferroelectrics, and frequency control*, vol. 65, no. 7, pp. 1078–1092, 2018.

[7] J. M. Boone, N. Shah, and T. R. Nelson, "A comprehensive analysis of DgNCTcoefficients for pendant-geometry cone-beam breast computed tomography," *Medical Physics*, vol. 31, no. 2, pp. 226–235, 2004.

[8] T. L. Szabo, *Diagnostic Ultrasound Imaging: Inside Out*. Academic Press series in biomedical engineering, Elsevier Science, 2004.

[9] F. J. Fahy, *Foundations of Engineering Acoustics*. Elsevier Science, 2000.

[10] American Piezo Ceramics, "Physical and Piezoelectric Properties of APC Materials."

[11] B. E. Treeby, J. Budisky, E. S. Wise, J. Jaros, and B. T. Cox, "Rapid calculation of acoustic fields from arbitrary continuous-wave sources," *The Journal of the Acoustical Society of America*, vol. 143, no. 1, pp. 529–537, 2018.

[12] B. E. Treeby and B. T. Cox, "k-Wave: MATLAB toolbox for the simulation and reconstruction of photoacoustic wave fields," *Journal of Biomedical Optics*, vol. 15, no. 2, p. 021314, 2010.

[13] Haifeng Wang, T. Ritter, Wenwu Cao, and K. K. Shung, "High frequency properties of passive materials for ultrasonic transducers," *IEEE Transactions on Ultrasonics, Ferroelectrics, and Frequency Control*, vol. 48, pp. 78–84, jan 2001.

[14] M. G. Grewe, T. R. Gururaja, T. R. Shrout, and R. E. Newnham, "Acoustic Properties of Particle/Polymer Composites for Ultrasonic Transducer Backing Applications," *IEEE Transactions on Ultrasonics, Ferroelectrics, and Frequency Control*, vol. 37, no. 6, pp. 506–514, 1990.

[15] B. Zeqiri, W. Scholl, and S. P. Robinson, "Measurement and testing of the acoustic properties of materials: A review," *Metrologia*, vol. 47, no. 2, 2010.

[16] S. W. Hughes, "Archimedes revisited: A faster, better, cheaper method of accurately measuring the volume of small objects," *Physics Education*, vol. 40, no. 5, pp. 468–474, 2005.

[17] A. Devaney and H. Levine, "Effective elastic parameters of random composites," *Applied Physics Letters*, vol. 37, no. 4, pp. 377–379, 1980.

[18] Y. Kim, A. D. Maxwell, T. L. Hall, Z. Xu, K. Lin, and C. A. Cain, "Rapid prototyping fabrication of focused ultrasound transducers," *IEEE Transactions on Ultrasonics, Ferroelectrics, and Frequency Control*, vol. 61, pp. 1559–1574, Sep. 2014.

[19] J. R. Morris and T. Wojcik, "Stencil printing of solder paste for fine-pitch surface mount assembly," *IEEE transactions on components, hybrids, and manufacturing technology*, vol. 14, no. 3, pp. 560–566, 1991.

[20] C. K. Aidun and N. G. Triantafillopoulos, *High-Speed Blade Coating*, pp. 637–672. Dordrecht: Springer Netherlands, 1997.

[21] N. V. Ruiter, M. Zapf, T. Hopp, R. Dapp, and H. Gemmeke, "Optimization of the aperture and the transducer characteristics of a 3d ultrasound computer tomography system," in *Medical Imaging 2014: Ultrasonic Imaging and Tomography*, vol. 9040, p. 90400U, International Society for Optics and Photonics, 2014.

[22] C. E. Baumgartner, D. M. Mills, R. S. Lewandowski, L. S. Smith, and D. G. Wildes, "Backing material for micromachined ultrasonic transducer devices," 2004.

[23] X. Zeng and R. J. McGough, "Evaluation of the angular spectrum approach for simulations of near-field pressures," *The Journal of the Acoustical Society of America*, vol. 123, no. 1, pp. 68–76, 2008.

A Low-cost Ultrasound Computed Tomography System using Diagnostic Linear Arrays

P. Patel[1], B. Cox[1], and B. Treeby[1]

[1] *University College London, London, United Kingdom, E-mail: zchapkp@ucl.ac.uk*

Abstract

Ultrasound Computed Tomography is clinical imaging modality whose potential to complement mammography as a non-ionising, economical breast screening or diagnosis tool has recently driven renewed interest and rapid development. Costly, custom-built detector arrays with a variety of geometries have been proposed including bowl-shaped, rings, and planar arrays. Here, we investigate whether a simple low-cost system based on diagnostic linear arrays can be used to perform USCT. A proof-of-concept system was constructed using clinical-grade linear arrays, an off-the-shelf data acquisition system, and in-house image reconstruction software.

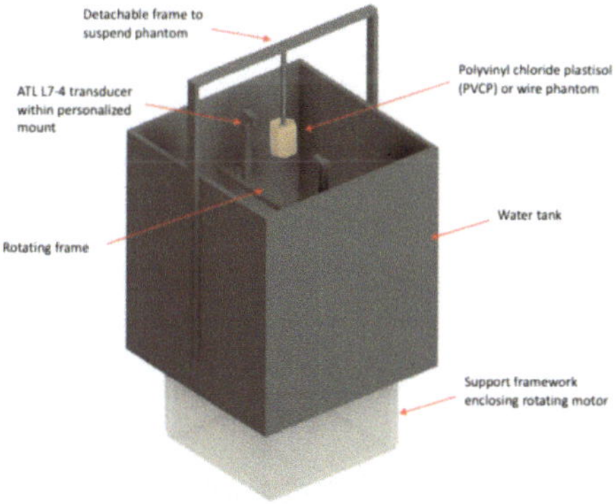

Figure 1: A schematic of the USCT.

The designed USCT system (see Figure 1) houses two linear array transducers, opposing each other, rotating on a frame about a phantom. The arrays fit to the frame via a detachable mount, allowing for an adjustable array separation and hence reducing the effects of diffraction for phantoms of varying dimensions. The main operating components are located beneath the water tank and hence it can be placed under a patient bed for clinical applications in the future. The entire system excluding the transducers and phantoms is designed from recycled plastic and stainless steel, making it economical and simple to manufacture. The probes are connected to a Verasonics ultrasound system which allows for independent channel and element control. Control of the rotation motor and scan sequencing was via Matlab. Four phantoms were scanned (two PVCP and two wire phantoms). At each rotation angle, a plane wave was emitted from an array and both the back-scattered and transmitted signals were detected. Time-of-flight and amplitude values were extracted from each measured time series and, and 2D images of the sound speed and attenuation were reconstructed using both straight- and bent-ray reconstruction algorithms.

The systems components and promising preliminary results are to be presented. This will include sound speed and attenuation maps from multiple signal processing and reconstruction algorithms, data from multiple scans with varying angles of rotation as well as a comparative analysis of the above scan parameters. The system characteristics (e.g. resolution) from the wire phantom will also be discussed.

Towards 3D Brain Imaging in Small Animals using Full-Waveform Inversion

T. Robins[1], C. Cueto[1], O. Calderon Agudo[1], L. Guasch[1], M. Warner[1], and M.-X. Tang[1]

[1] *Imperial College London, London, United Kingdom, E-mail: tcr11@ic.ac.uk*

Abstract

Full-waveform inversion (FWI) is an iterative image-reconstruction technique used in exploration seismology to produce high-quality tomographic images for acoustic properties of the subsurface by minimising the misfit between acquired and modelled data FWI has also been shown to have some exciting applications in ultrasound imaging of soft tissue, including breast cancer diagno- sis, however the method has yet to be applied to brain-imaging in small animal models where hard tissue is present and 3D reconstruction is required. Here we demonstrate the feasibility of 3D small-animal transcranial tomography using FWI in silico and implement our FWI algorithm to reconstruct an in vitro acoustic brain phantom in 2D using data acquired with independently translated medical transducers.

In this study a pair of 96-element linear probes (P4-1, ATL) were each assigned a set of 3-axis motors for independent translation, allowing for a larger imaging region and for 3D acquisitions to be performed. To test this experimental setup, an in silico mouse brain and skull acoustic model was created and placed between two 1152-element plane arrays, each representing the translation of a linear array over 12 positions (with 2.5 mm spacing), as shown in Fig1.a. A synthetic 3D dataset of this scene was produced so that the velocity model could be reconstructed using a robust 3D FWI algorithm. To mitigate cycle skipping due to the skull tissue, a starting model consisting of the mouse skull and homogenous water (1500 ms^{-1}) was used. To demonstrate this method using real data, we acquired a 2D dataset of a half-scale 2.5D brain phantom made using a polyvinyl alcohol cryogel to mimic brain tissue. By translating both probes vertically when acquiring data, a two-plane array consisting of 960 elements was formed, from which a subset of 240 were selected as sources. Element delays and positions were calibrated by time of flight optimisation using a watershot dataset and the velocity model was reconstructed using our 2D FWI algorithm.

Fig1.b shows FWI results of the mouse brain and skull model. When compared to the ground truth, a relative error of 0.110 % has been achieved (starting model relative error = 2.562 %). In Fig1.c-d the photo of the brain phantom and the reconstructed velocity model can be seen to closely match. In future work we will introduce rotation to further improve our transducer scanning sequence and apply our method to image ex-vivo brain and skull tissue.

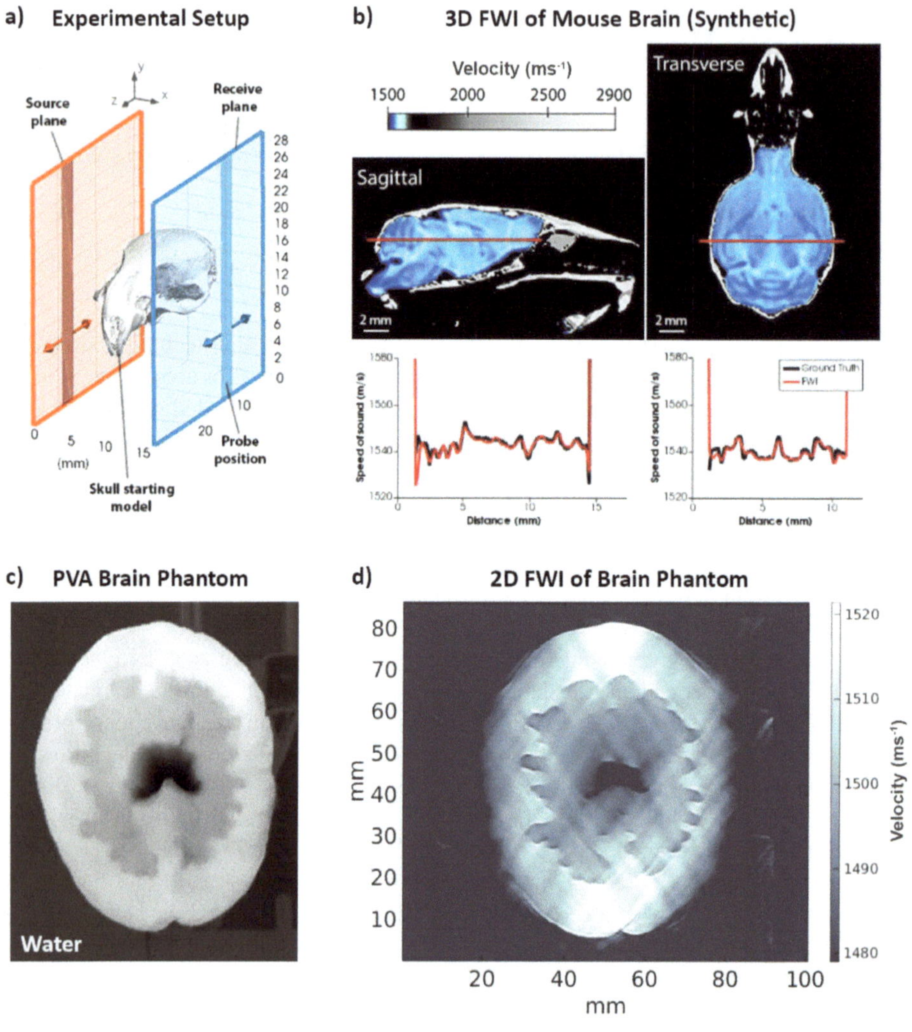

Figure 1: **a)** Experimental setup of 3D synthetic mouse brain model, **b)** Saggital and transverse views of velocity model recovered using FWI with velocity samples (red) plotted against ground truth, **c)** Photo of brain phantom, **d)** Brain phantom velocity model recovered using FWI.

The Efficiency of an All-reflective Omnidirectional Illumination for Photoacoustic Tomography with a Ring Ultrasound Transducer

S. S. Alshahrani[1], N. Alijabbari[1], A. Pattyn[1], and M. Mehrmohammadi[1,2,3]

[1] *Wayne State University, Department of Biomedical Engineering, Detroit, USA*
E-mail: ep7200@wayne.edu
[2] *Wayne State University, Department of Electrical and Computer Engineering, Detroit, USA*
[3] *Barbara Ann Karmanos Cancer Institute, Detroit, USA*

Abstract

Breast cancer is a common cancer and a major health concern affecting the lives of many women worldwide. According to the World Health Organization (WHO) breast cancer has an estimated annual incidence rate of 2.1 million and is responsible for about 15% of all cancer related deaths around the world. We have previously introduced a novel photoacoustic tomography (PAT) system for breast cancer imaging, in which the object is illuminated omnidirectionally by using a cone and a conical ring mirrors. A ring ultrasound (US) transducer is used to acquire PA signals and to form the PAT images. The design of the system is intended to acquire data from entirety of a pendant breast up to the chest wall, without compressing the tissue, thus allowing for deeper detection of endogenous chromophores associated with breast cancer.

This work presents the characterization of the full-ring illumination / acquisition system through three different studies. In the first study, three different illumination methods, full-ring, diffused-beam, and point source illumination, and their efficacy for PAT imaging are compared. The results indicate that the full-ring illumination method is capable of providing a more uniform fluence irrespective of the vertical depth of the cross-section imaged, while the point source and diffused illumination methods provide a higher fluence at regions closer to the point of entry, which diminishes with depth. The omnidirectional ring illumination can provide a more uniform illumination pattern within the tissue and enable PAT imaging at deeper distances in the targeted cross-sectional area. In the second study, a set of experiments were conducted to determine the optimum position of ring-illumination with respect to the position of the acoustic detectors to achieve the highest signal-to-noise ratio. The last study was performed on breast- mimicking phantom, made out of human fat and embedded blood inclusions, to truly identify the capabilities of the developed PAT system and specifically the achievable in-plane penetration depth. The results demonstrate that the system is capable of imaging blood up to a cross-sectional depth of 30 mm. These three studies demonstrate the utility and advantages of the full ring-illumination system in imaging breast mimicking objects and paves the way for future development of the system towards a clinically translatable breast PAT scanner and can be part of a future diagnostic system.

Clinical Studies

Tissue Sound Speed: A Novel Imaging Biomarker for Measuring Tamoxifen Response

M. Sak[1], N. Duric[1,2], P. Littrup[1], and G. Gierach[3]

[1] *Delphinus Medical Technologies, Inc, E-mail: msak@delphinusmt.com*
[2] *Karmanos Cancer Institute*
[3] *National Cancer Institute*

Abstract

Purpose
Studies have shown that a decrease in mammographic percent density (MPD) or lowering of background parenchymal enhancement (BPE) on MRI after initiation of tamoxifen therapy predicts a favorable response in the preventive or adjuvant settings. However, performing serial mammograms poses radiation concerns, while serial MRIs carry high cost as well as risk of multiple Gadolinium doses. Previous studies have shown that tissue sound speed, derived from whole breast ultrasound tomography measurements, is a surrogate biomarker of MPD. Ultrasound is ideal for performing serial measurements because it is fast and poses almost no risks. The purpose of this study was to evaluate repeated measures of the sound speed biomarker at 3, 6 and 12-months following tamoxifen initiation.

Method and Materials
We performed a case-control comparison involving 74 participants referred by a health professional to undergo tamoxifen therapy (cases) and 150 matched participants with no history of breast cancer (controls). The cases were scanned with ultrasound tomography at baseline (i.e. before start of tamoxifen therapy), and then at 1-3, 3-6 and 12 months after tamoxifen initiation. Controls were scanned at baseline and 12 months. In the case group, sound speed was measured pre-treatment in the contralateral breast to avoid potential influences of tumor-related changes on density. In the control group, a single randomized breast was scanned. A paired t-test was used to assess differences in sound speed and MPD over time and compared to baseline.

Results
There was a steady decline in sound speed over the 12-month period for women undergoing tamoxifen therapy (mean: -2.6 m/s, $p = 0.005$). Furthermore, significant sound speed reductions were observed as early as 3-6 months after tamoxifen initiation (mean: -2.3 m/s, $p = 0.007$); Figure 1. In contrast, the controls demonstrated no significant change in sound speed over a 12-month period.

Conclusion
Breast sound speed decreases rapidly after tamoxifen initiation; further studies are needed to assess whether this can predict clinical response.

Clinical Relevance / Application
Ultrasound tomography may have utility in monitoring breast sound speed change as a potential biomarker of clinical tamoxifen response.

Keywords: Breast Density, Tamoxifen, UST Soundspeed

1 Introduction

Tamoxifen is a selective estrogen receptor modulator (SERM) that has been shown to reduce the incidence of breast cancer in women by up to 50% and is therefore commonly used as a breast cancer preventative agent [1,2]. The reduction in breast cancer risk is also associated with a decrease in breast density, particularly in premenopausal women. Favourable response in either the preventive or adjuvant settings have been shown to be associated with either decreases in mammographic percent density (MPD) or lowering of background parenchymal enhancement (BPE) [3,4].

The effects of decreased breast density (BD) are the most pronounced with the first 12-18 months [3]. Yet both mammography and MRI have shortcomings in regard to the repeated serial scans that would be needed to accurately track response to Tamoxifen on a time frame shorter than this. The use of radiation in mammography along with both the use of Gadolinium and the high cost of MRI would limit the period between repeated scans. Increased monitoring early on in treatment could therefore improve patient outcomes by identifying patients who are or aren't responding and adjusting course appropriately.

Studies have shown that average tissue sound speed, a factor derived from whole breast ultrasound tomography (UST) scans, is strongly correlated with mammographic breast density [5,6]. Ultrasound tomography overcomes many of the issues associated with mammography and MRI. Namely, UST is non-ionizing, does not use contrast and is much less expensive than MRI. This makes UST imaging an ideal modality for performing serial measurements.The aim of this work is to perform a preliminary evaluation on the repeated measures of UST average sound speed at short, regular intervals following Tamoxifen initiation.

2 Methods

2.1 Patient Recruitment

A total of 73 patients beginning treatment with Tamoxifen, either due to a history of breast cancer or those with a higher risk of developing breast cancer were selected to be part of this trial. These "case" patients were referred by health professionals mainly at the Karmanos Cancer Institute (KCI) in Detroit. A total of 140 "control" patients with negative mammographic screens who were not receiving Tamoxifen were also enrolled. These control patients were age and race matched with the case patients. The study was performed under an IRB approved protocol and was HIPPA compliant.

All case patients received a mammogram and UST scan at baseline (prior to start of Tamoxifen therapy) followed by repeated UST scans at 1-3, 3-6 and 12 months post baseline along with a repeat mammogram at 12 months in accordance with yearly screening protocols. All control patients received their two normal screening mammograms 12 months apart combined with UST scans at the same time. A single randomly chosen breast was analyzed for each control patient and each case patient with no history of breast cancer. For those case patients with recently diagnosed breast cancer, the contralateral breast chosen for analysis.

2.2 Breast Density Measurements

Mammographic percent density (MPD) was analyzed by one expert reader using the CUMULUS 4 Software. This software uses a semi-automated algorithm where the reader manually selects the breast and a threshold between the dense and fatty tissue. The software then measures the amounts of dense and total areas on the mammogram which are then used to calculate the percent density.

UST images were created using a prototype device at the Karmanos Cancer Institute that could produce between 40-100 sound speed images per breast. Whole volume averaged sound speed (VASS) was measured using the public domain software package ImageJ. Because the soundspeed of the background water bath is intermediate to that of the normal breast tissue, segmentation of the breast tissue was done using a semi-automated method that approximated the breast in each slice as an ellipse. After the segmentation was completed, the remaining pixels could be averaged to produce the VASS.

3 Results

There was a decline in VASS for case patients compared to control patients. On average, the VASS in the case group decreased by 2.6 m/s after 12 months compared to the baseline sound speed. This change was statistically significant as a paired t-test was performed and a p value of $p = 0.005$ was measured. The change in sound speed could be detected as early as 3-6 months post the initiation of Tamoxifen as an average decrease in sound speed of 2.3 m/s ($p = 0.007$) was noted. A slight increase in VASS of 0.3 m/s ($p = 0.5$) was noted for controls. These results can be visualized in Figure 1.

The changes in MPD were not as apparent. For cases, there was a decrease in average MPD of 2.3% after 12 months, but this was not statistically significant (paired t-test $p = 0.11$). For controls, there was also a decrease of 1.8% and this was statistically significant ($p = 0.045$). These results can be seen in Figure 2. The changes in VASS and MPD are summarized in Table 1.

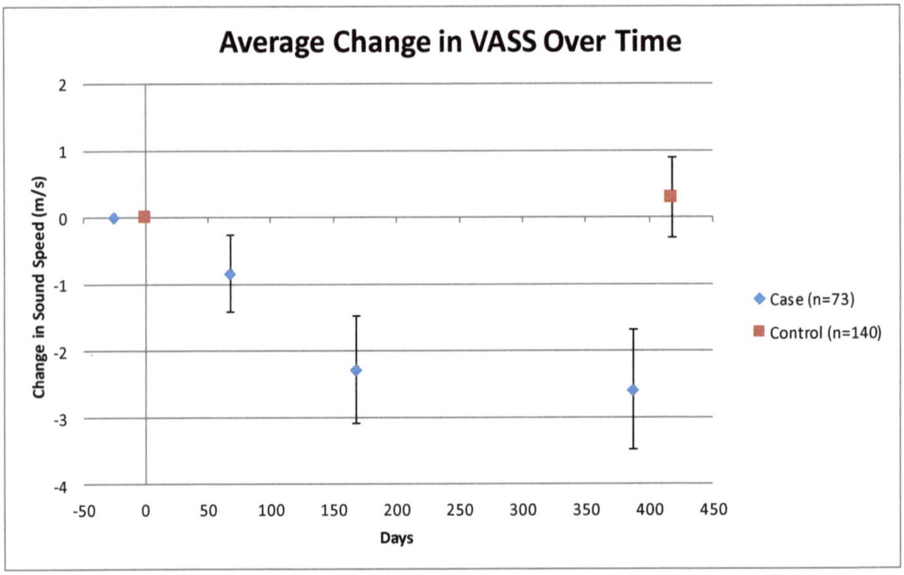

Figure 1: Average change in VASS for both cases and controls.

4 Discussion

When using UST density measures, average changes in breast density could be detected in the case group much sooner than could be detected when using mammographic measures. The differences when measured by UST were also statistically significant compared to the

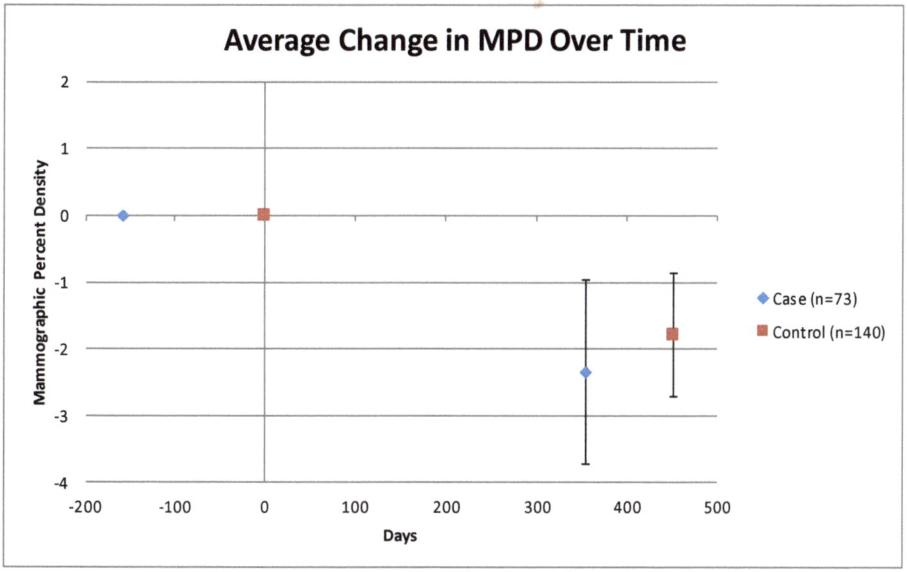

Figure 2: Average change in MPD for both cases and controls.

differences measured by mammography. This suggests that VASS may be a more sensitive measure of breast density and breast cancer risk than mammography. Previous results using only the baseline data suggests that UST sound speed potentially showed stronger breast cancer risk assessment than mammography and these findings may support that result [7].

There were inconsistencies in the density measures for the control group with UST sound speed slightly increasing but MPD showing a statistically significant decrease over the 12 months. Breast density is known to decrease with increasing age, so some small decrease in density is expected in both the case and control group.

Group	Density Measure Relative to Baseline	Average Change	p-value*
Case	1-3 Months VASS	-0.8	0.13
	3-6 Months VASS	-2.3	0.007
	12 Months VASS	-2.6	0.005
	12 Months MPD	-2.3	0.11
Control	12 Months VASS	0.3	0.5
	12 Months MPD	-1.8	0.045

* p-values are calculated from the paired t-test between the measured and the baseline values

Table 1: Summary of average changes in UST and mammographic density over time

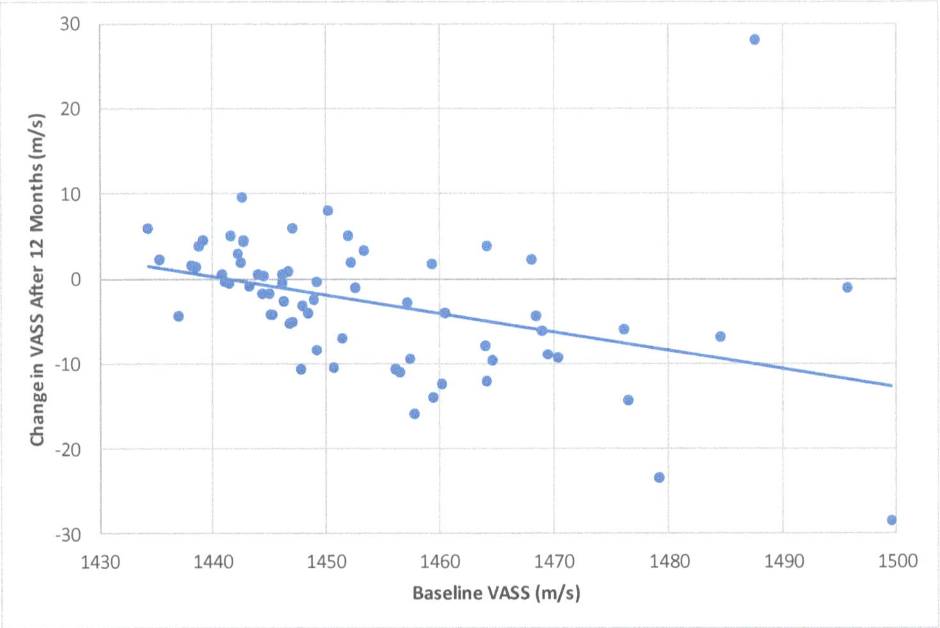

Figure 3: Plot of Baseline VASS vs the change in VASS after 12 months for all Cases. The Spearman correlation coefficient was r_s = -0.535

The average breast density of the patients in this study skews towards being less dense. Previous mammographic analysis of Tamoxifen's effect on breast density showed significant decreases in density in cases compared to controls. However, for those studies, the average baseline breast density was greater than 40% [3] while in this study, the baseline MPD for cases was 25% and 22% for controls. Since breast density started low, there was less of a potential decrease in breast density available to be made.

Patients with higher baseline breast density (measured with either VASS or MPD) generally showed the greatest decrease in breast density. There were moderate and negative correlations between the baseline density and the final change in density at 12 months. Figure 3 shows a plot of the baseline VASS and the change in sound speed after 12 months for the cases. A Spearman correlation of r_s = -0.535 was measured for this and Table 2 lists the Spearman correlations for the other different possible groupings. Despite the limitations of the dataset, VASS appears to be a more sensitive measure of breast density and changes to breast density than mammography. However, further analysis is still required to determine whether these changes can be used to predict clinical response.

Group	Density Measure Relative to Baseline	Spearman Coefficient	p-value*
Case	1-3 Months VASS	-0.307	0.01
	3-6 Months VASS	-0.345	0.004
	12 Months VASS	-0.535	< 0.001
	12 Months MPD	-0.461	< 0.001
Control	12 Months VASS	-0.219	0.007
	12 Months MPD	-0.378	< 0.001

Table 2: Correlations between baseline density measure and change in density measure

5 Conclusion

Ultrasound tomography may have utility in serially monitoring sound speed change as a potential biomarker of clinical Tamoxifen response. Changes in breast density may be more sensitively detected using VASS than by using MPD. Further analysis is still required to know if these changes can be used to predict clinical response.

References

[1] J. Brisson, B. Brisson, G. Coté, E. Maunsell, S. Bérubé, and J. Robert, Tamoxifen and Mammographic Breast Densities, Cancer Epidemiol. Biomarkers Prev., vol. 9, no. 9, pp. 911–915, 2000.

[2] E. B. C. T. C. Group, Tamoxifen for early breast cancer: an overview of the randomised trials, The Lancet, vol. 351, no. 9114. Early Breast Cancer Trialists' Collaborative Group, pp. 1451–1467, 1998, doi: http://dx.doi.org/10.1016/S0140-6736(97)11423-4.

[3] J. Cuzick et al., Tamoxifen-Induced Reduction in Mammographic Density and Breast Cancer Risk Reduction: A Nested Case-Control Study, J. Natl. Cancer Inst., vol. 103, no. 9, pp. 744–752, 2011.

[4] N. F. Boyd, Tamoxifen, Mammographic Density, and Breast Cancer Prevention, J. Natl. Cancer Inst., vol. 103, no. 9, pp. 704–705, 2011.

[5] N. Duric et al., Breast density measurements with ultrasound tomography: A comparison with film and digital mammography, Med. Phys., vol. 40, no. 1, pp. 13501–13512, 2013.

[6] M. Sak et al., Using Speed of Sound Imaging to Characterize Breast Density, Ultrasound Med. Biol., vol. 43, no. 1, pp. 91–103, Jan. 2017, doi: 10.1016/J.ULTRASMEDBIO.2016.08.021.

[7] N. Duric et al., Using Whole Breast Ultrasound Tomography to Improve Breast Cancer Risk Assessment: A Novel Risk Factor Based on the Quantitative Tissue Property of Sound Speed, J. Clin. Med., vol. 9, no. 2, p. 367, Jan. 2020, doi: 10.3390/jcm9020367.

Tissue Sound Speed from Ultrasound Tomography as a Biomarker of Breast Cancer Risk: A Review of Recent Advances

N. Duric[1,2], P. Littrup[1], M. Sak[1], and G. Gierach[3]

[1] *Delphinus Medical Technologies, Inc, E-mail: nduric@delphinusmt.com*
[2] *Karmanos Cancer Institute*
[3] *National Institutes of Health*

Abstract

Increased mammographic density is a strong population-based risk factor for breast cancer. However, the inclusion of breast density in clinical risk models leads to only a modest improvement in individual risk prediction. Ultrasound Tomography (UST) is an emerging breast imaging technique that has the potential to overcome some of mammography's limitations as a predictor of breast cancer risk.

We review the methods and results of previous UST studies investigating the use of tissue sound speed to characterize breast density and breast cancer risk. A number of different UST methods have yielded strong correlations between speed and breast density with Spearman correlation coefficients ranging from 0.71 to 0.96. and in some cases, even stronger correlations between sound speed and MRI measures of breast density with correlation coefficients of up to 0.96. In one breast cancer risk study, elevated sound speed was significantly associated with increased breast cancer risk (odds ratios (OR) per quartile=1.79, 95%, CI: 1.30, 2.48; p_{trend}=0.0004). In contrast, Mammographic density was associated with elevated breast cancer risk compared to controls, although the trend was not statistically significant (OR per quartile=1.28, 95%CI: 0.95, 1.73; p_{trend}=0.10). The OR-trend for sound speed was significantly different from that observed with mammography (p=0.01).

Recent clinical studies have demonstrated that sound speed not only corelates with mammographic density but is also related to breast cancer risk. UST has the potential to provide a more accurate, non-ionizing method for assessing breast density and breast cancer risk.

Keywords: Ultrasound Tomography, sound speed, breast cancer risk

1 Introduction

Many clinical studies have shown that breast density (BD), as measured by mammography, is a strong risk factor for breast cancer [1]. At the same time, women with high breast density are the least likely to benefit from mammographic screening where sensitivity can be less than 50% [2]. Thus, this high-risk population that could benefit the most from breast cancer screening, is, in fact, underserved.

Mammographic studies have established a relationship between mammographic density (MD) and breast cancer (BC) risk. When using MD quartiles, the women in the highest density quartile have a 3 to 6-fold increase in BC risk compared to women in the lowest density quartile [2-7]. While this risk factor is strong, its inclusion in clinical risk models has resulted in limited additional benefit. While a typical clinical risk model (e.g. Gail or Tyrer-Cuzick) predicts individual risk correctly 60% of the time, inclusion of MD raised the prediction accuracy to 61% - 64% [8]. Thus, MD remains a useful marker for characterizing populations but has only modest value in predicting individual risk. There is therefore a need for a mor sensitive measure of breast cancer risk.

The MD measurements are limited by several factors. First, mammography measures MD from 1 or more 2D projections of a highly compressed 3D volume. Any such projection reduces the information contained in the original undistorted volume. Secondly, it requires segmentation of mammograms to determine the percent of dense tissue contained within the beast. Such segmentation requires subjective or otherwise user defined thresholding. Thirdly, digital mammography is optimized for mass detection, not density measurements. And finally, MD is not a quantitative measure in the true sense. It does not yield pixel values that can tied to an external standard. These limitations form a likely set of reasons why MD's contribution to clinical risk models is modest. An alternate imaging approach that mitigates one or more of these limitations has the potential to improve upon MD as a BD risk marker.

Ultrasound Tomography (UST) is an emerging breast imaging technique that is being developed to overcome some of mammography's limitations and has the potential to aid in the screening of women with dense breasts [9-15]. In this paper, we review its role in assessing breast density and breast cancer risk. Unlike mammography, UST is a 3D imaging technique that does not compress the breast and creates a 3D imaging set so that information in the natural breast volume is preserved. Previous UST studies have shown that the speed of sound, a tissue property that is measured by UST, is closely linked to breast density [16-21]. In fact, some studies have suggested that sound speed is directly linked to physical tissue density [19].

2 Methods

Measuring breast density. Previous UST studies relating to BD were carried in cohorts of women undergoing mammographic exams with UST scans added as part of the studies [16-21]. These studies were IRB approved and HIPPA compliant. UST scans were performed with the breast immersed in water with one or more transducers moving through the water to acquire the signals needed to perform a tomographic inversion.

The measurement of MD is usually carried out using segmentation software such Cumulus or Libra, which require some degree of subjectivity to carry out. In some studies, only the radiologist assessment of the BIRADS breast density category (a,b,c,d) is used. Automated methods, such as Quantra and Volpara have been introduced in recent years to facilitate the MD measurement and to attempt volumetric estimates by taking into account the thickness of the breast under compression.

As noted above, UST studies measure sound speed as a surrogate of BD. Extraction of sound speed information, as it relates to BD, has been carried out in several ways.

(i) Thresholding has been used to generate estimates of percent densities by setting a cut point in sound speed above which is considered to be dense tissue. The advantage of this approach is that it is simple to implement. The downside is that there is no standard or universal agreement on what that threshold should be since no prospective clinical trials have been carried out to determine the optimal value.

(ii) Segmentation techniques, such as k-means clustering have been used to determine percent densities by identifying clusters of pixels that represent dense (high sound speed) tissue. This approach has the advantage that it uses a standard recognized clustering technique and does not require an arbitrary or subjective threshold value. It has the disadvantage that is more computationally complex and like the thresholding method, it yields percentages only.

(iii) The volume averaged sound speed (VASS) of the breast is a parameter that provides a true quantitative value by characterizing the breast with a single number that is tied to an external standard, namely the speed of sound in tissue, usually expressed in meters/second. An additional advantage is that it is purely objective and simple to calculate. Its main drawback is that it is more difficult to relate to standard MD measures. However, as shown below it does not need to be tied to MD to demonstrate its value as a BC risk factor.

Risk Assessment. So far, only one study has been carried out for assessing UST sound speed as an independent risk factor for breast cancer [22]. In this study, breast cancer risk associated with VASS and MD was evaluated in a case-control study involving 59 participants with recent breast cancer diagnoses (cases, aged 30-70 years) and 150 participants with no history

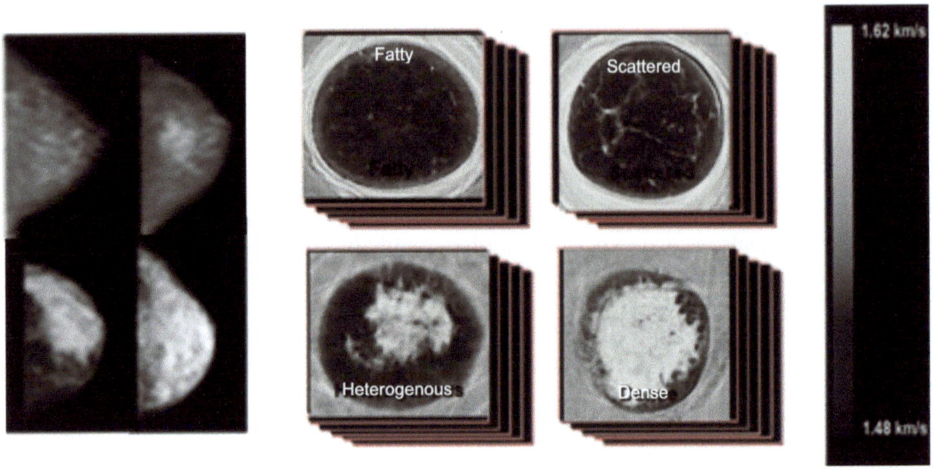

Figure 1: Examples of Mammograms (left) and UST images of sound speed (right) shown for the 4 BIRADS BD categories.

of breast cancer (controls), who were matched to cases on age, race, and menopausal status. The cases and controls were imaged with both ultrasound tomography (UST) and mammography. In cases, breast density was measured pre-treatment in the contralateral breast to avoid potential influences of tumor-related changes on MPD or sound speed. In controls, a randomly selected breast was imaged. The ultrasound tomography images were used to estimate the volume averaged sound speed of the breast, and the Cumulus software package was applied to mammograms to determine MPD. Odds Ratios (ORs) adjusted for matching factors and 95% Confidence Intervals (CIs) were calculated for the relation of quartiles of MPD and VASS with breast cancer risk.

3 Results

Breast density. Breast density. The three UST methods described above have yielded strong correlations between speed and MD with Spearman correlation coefficients ranging from 0.71 to 0.96 [16-21]. and in some cases, even stronger correlations between sound speed and MRI measures of BD with correlation coefficients of 0.96 [18]. Figure 1 shows examples of cases in the four different BIRADS density categories, showing the correlation between MD and UST sound speed.

Breast Cancer Risk. In the one BC risk study, elevated VASS was significantly associated with increased breast cancer risk in a dose-response fashion (OR per quartile=1.79, 95%CI: 1.30, 2.48; p_{trend}=0.0004) (Figure 2). In contrast, MPD was associated with elevated breast cancer risk compared to controls, consistent with previous studies, although the trend did not reach statistical significance (OR per quartile=1.28, 95%CI: 0.95, 1.73; p_{trend}=0.10). The OR-trend for VASS was statistically significantly different from that observed for MPD (p=0.01). Figure 2 illustrates the OR's of MD vs VASS. The ratio of highest to lowest density quartiles (Q4/Q1)was found to be 8.6 for VASS vs 1.76 for MD.

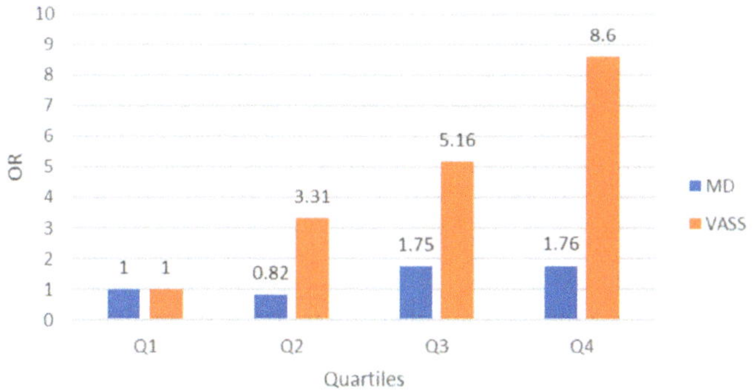

Figure 2: Histogram of ORs by quartiles. MD data shown in blue, VASS data in orange.

4 Discussion

Recent clinical studies have demonstrated that sound speed not only corelates with MD but is also related to BC risk. We present two possible interpretation of these relations.

Sound speed is a proxy of MD. Since MD is known to be associated with BC risk and since sound speed itself is associated with BD, it stands to reason that VASS would be associated with BC risk. While the biological cause of the increased BC risk is not well understood, the driver of the relationship appears to be the dense tissue content of the breast.

Under this scenario, the stronger association of VASS with BC risk can be explained by the fact that UST mitigates some of the limitations of mammography, as described in the Introduction and provides a more comprehensive picture of the volumetric distribution of dense tissue. Furthermore, sound speed is a measure of physical density of tissue suggesting that it is the physical density, not MD, that drives the observed relationship between VASS and BC risk.

Sound speed is an independent marker of BC risk. The relationship between VASS and BC risk was observed independently of MD. In other words, it stands as a risk factor on its own without a need to tie it to breast density. Thus, it is plausible that sound speed carries additional biological information on BC risk beyond breast density.

The much stronger association of VASS with BC risk compared to MD (Q4/Q1 = 8.6 vs 1.76) suggests an almost 5-fold improvement in risk stratification. Such a large increase suggests that VASS may be a sensitive measure of an underlying biological factor that has yet to be identified.

In a clinical setting, UST would provide relatively comfortable exams without the use of ionizing radiation and at relatively low cost. In that context, determining VASS as a risk factor may be viable. Furthermore, the absence of radiation would make this risk assessment available to women for all ages. Interventions for women at high risk are most effective when performed at early ages so the use of UST for young women (below screening age) would provide timely data, guiding interventions that reduce their risk.

Future studies are needed to validate the findings presented above, investigate the underlying cause of the association between VASS and BC risk and to test the VASS parameter in a clinical risk model so that its impact om individual risk prediction can be assessed. Such studies may lead to the use of VASS as a biomarker of risk, as defined by the Food and Drug Administration (FDA) under its “context of use” guidelines. Finally, future studies may investigate additional parameters that UST can measure, such as tissue attenuation and stiffness that can be added to risk models to improve their accuracy.

5 Conclusion

Elevated breast density strongly increases breast cancer risk. UST has the potential to provide a more accurate, non-ionizing method for assessing breast density and its associated breast cancer risk.

References

[1] McCormack VA, dos Santos Silva I (2006) Breast density and parenchymal patterns as markers of breast cancer risk: a meta-analysis. Cancer epidemiology, biomarkers & prevention : a publication of the American Association for Cancer Research, cosponsored by the American Society of Preventive Oncology 15 (6):1159-1169.

[2] Wolfe JN (1976) Risk for breast cancer development determined by mammographic parenchymal pattern. Cancer 37 (5):2486-2492

[3] Boyd NF, Guo H, Martin LJ, Sun L, Stone J, Fishell E, Jong RA, Hislop G, Chiarelli A, Minkin S, Yaffe MJ (2007) Mammographic density and the risk and detection of breast cancer. The New England journal of medicine 356 (3):227-236.

[4] Boyd N, Guo H, Martin L, et al. 2007. Mammographic density and the risk and detection of breast cancer. N Engl J Med, 356(3), 227- 236. PMID: 17229950.

[5] Byng JW, Boyd NF, Fishell E, Jong RA, Yaffe MJ (1994) The quantitative analysis of mammographic densities. Physics in medicine and biology 39 (10):1629-1638

[6] Boyd N, Martin L, Gunasekara A, Melnichouk O, Maudsley G, Peressotti C, Yaffe M, Minkin S (2009) Mammographic density and breast cancer risk: evaluation of a novel method of measuring breast tissue volumes. Cancer epidemiology, biomarkers & prevention : a publication of the American Association for Cancer Research, cosponsored by the American Society of Preventive Oncology 18 (6):1754-1762.

[7] Yaffe MJ, Boone JM, Packard N, Alonzo-Proulx O, Huang SY, Peressotti CL, Al-Mayah A, Brock K (2009) The myth of the 50-50 breast. Medical physics 36 (12):5437-5443

[8] Amir E, Freedman OC, Seruga B, and Evans DG. 2010. Assessing women at high risk of breast cancer: a review of risk assessment models. J Natl Cancer Inst, 102(10), 680–691. PMID: 20427433 (Cummings et al., 2009).

[9] Carson PL, Meyer CR, Scherzinger AL, Oughton TV. Breast imaging in coronal planes with simultaneous pulse echo and transmission ultrasound. Science 1981, Dec 4;214(4525):1141-3.

[10] Andre MP, Janee HS, Martin PJ, Otto GP, Spivey BA, Palmer DA, "High-speed data acquisition in a diffraction tomography system employing large-scale toroidal arrays," International Journal of Imaging Systems and Technology 1997;Vol. 8, Issue 1:137-147.

[11] Johnson SA, Borup DT, Wiskin JW, Natterer F, Wuebbling F, Zhang Y, Olsen C. Apparatus and Method for Imaging with Wavefields using Inverse Scattering Techniques. United States Patent 6,005,916 (1999).

[12] Marmarelis, V.Z., Kim, T., Shehada, R.E. Proceedings of the SPIE: Medical Imaging 2003; San Diego, California; Feb. 23-28, 2002. Ultrasonic Imaging and Signal Processing – Paper 5035-6.

[13] Liu, D.-L., and Waag, R. C. "Propagation and backpropagation for ultrasonic wavefront design," IEEE Trans. on Ultras. Ferro. and Freq. Contr. 44(1):1-13 (1997).

[14] Duric N, Littrup PJ, Poulo L, et al. Detection of breast cancer with ultrasound tomography: First results with the Computed Ultrasound Risk Evaluation (UST) prototype. Med. Phys. 2007; 34: 773-785

[15] Nicole V. Ruiter, Georg Göbel, Lutz Berger, Michael Zapf and Hartmut Gemmeke, "Realization of an optimized 3D USCT", Proc. SPIE 7968, 796805 (2011)

[16] Glide-Hurst CK, Duric N, Littrup P (2008) Volumetric breast density evaluation from ultrasound tomography images. Medical physics 35 (9):3988-3997

[17] Sak, M., N. Duric, P. Littrup, L. Bey-Knight, H. Ali, P. Vallieres, M.E. Sherman, and G.L. Gierach, Using Speed of Sound Imaging to Characterize Breast Density. Ultrasound Med Biol, 2017. 43(1): 91-103.

[18] O'Flynn, E.A.M., J. Fromageau, A.E. Ledger, A. Messa, A. D'Aquino, M.J. Schoemaker, M. Schmidt, N. Duric, A.J. Swerdlow, and J.C. Bamber, Ultrasound Tomography Evaluation of Breast Density: A Comparison With Noncontrast Magnetic Resonance Imaging. Invest Radiol, 2017. 52(6): 343-348.

[19] Duric, N., N. Boyd, P. Littrup, M. Sak, L. Myc, C. Li, E. West, S. Minkin, L. Martin, M. Yaffe, S. Schmidt, M. Faiz, J. Shen, O. Melnichouk, Q. Li, and T. Albrecht, Breast

density measurements with ultrasound tomography: a comparison with film and digital mammography. Med Phys, 2013. 40(1): 013501.

[20] Wiskin, J., Malik, B., Natesan, R. and Lenox, M. Quantitative assessment of breast density using transmission ultrasound tomography, Med. Phys. 46(6), 2610–2620 (2019).

[21] Khodr, Z.G., Sak, M.A., Pfeiffer, R.M., Duric, N., Littrup, P., Bey-Knight, L., Ali, H., Vallieres, P., Sherman, M.E. and Gierach, G.L., 2015. Determinants of the reliability of ultrasound tomography sound speed estimates as a surrogate for volumetric breast density. Medical physics, 42(10), pp.5671-5678.

[22] Duric, N., Sak, M., Fan, S., Pfeiffer, R.M., Littrup, P.J., Simon, M.S., Gorski, D.H., Ali, H., Purrington, K.S., Brem, R.F., Sherman, M.E. and Gierach, G.L., 2020. Using Whole Breast Ultrasound Tomography to Improve Breast Cancer Risk Assessment: A Novel Risk Factor Based on the Quantitative Tissue Property of Sound Speed. Journal of Clinical Medicine, 9(2), p.367.

Breast Cancer Development at the Fat-Gland Interface (FGI): Importance of Coronal Imaging and Ultrasound Tomography

P. J. Littrup[1-4], N. Duric[2,3], M. Sak[3], and R. Brem[4]

[1] *Prof. of Radiology, Wayne State University, Detroit, MI, USA, E-mail: plittrup@delphinusmt.com*
[2] *Prof. of Oncology, Wayne State University, Detroit, MI, USA*
[3] *Delphinus Medical Technologies, Inc., Novi, MI, USA*
[4] *George Washington University School of Medicine & Health Sciences, D.C., USA*

Abstract

Objective

We evaluated the locations of cancer and benign masses in relation to the fat-gland interface and anatomic quadrant locations with ultrasound tomography.

Methods

A total of 602 breast masses were evaluated from a clinical data set corresponding to a pre-trial arm of a multicenter study evaluating UST. Anatomic breast quadrants also noted for each tumor, and the fat-gland interface was considered the site of origin if at least 1/8 of the mass circumference abutted fat. In a subset of 298 masses, a hand-traced region-of-interest allowed quantitative sound speed and percent density comparisons of tumor and peritumoral margins.

Results

Cancers were located at the fat-gland interface in 93.4% (169/181) of cases, and were completely surrounded by fat in 5.5% (10/181) or parenchyma in 1.1% (2/21)($p<0.001$) of cases.

Moreover, 56.1% (97/173) of cysts and 25.0% (47/188) of fibroadenomas were fully surrounded by dense tissue, significantly more than cancers (1.1%)($p<0.001$). The upper outer quadrant was the most common location of all masses, but no significant differences between mass were noted ($p=0.06$). Quantitative sound speed values showed significantly greater amount of fat surrounding cancers than fibroadenomas or cysts ($p<0.001$), while percent density showed that most cysts were completely surrounded by dense tissue and cancers were surrounded by both dense and fatty tissues.

Conclusion
Both the qualitative tumor location and quantitative peritumoral data corroborate the preferential location of cancers at the FGI, compared to benign masses or anatomic quadrants. This study supports possible integration into clinical practice with future dense breast screening by UST.

Keywords: breast cancer, fat-glandular interface, ultrasound tomography

1 Introduction

1.1 Breast Imaging and the Fat-Glandular Interface (FGI)

Imaging of breast cancer development is limited to macroscopic findings (e.g., >5 mm) by each modality. 3D localizations of cancer origin according the dominant breast tissues of denser fibroglandular parenchyma or fat are not commonly considered. Common mammographic teaching has emphasize greater visual attention to anatomic regions of historically higher cancer incidence [1,2], such as the upper outer quadrants having its greater parenchymal content, or epithelial distribution [2]. The dominant signs of breast cancer by mammography relate to a focal mass/asymmetry, a cluster of microcalcifications and/or architectural distortion.

Cancers newly diagnosed by mammography were seen as a mass/asymmetry in over 2/3 of cases, by calcifications in 29% and architectural distortion in only 4% [3]. Masses or asymmetries seen by mammography are generally then evaluated by ultrasound, which predominantly characterizes it as a cyst or solid mass for follow-up or biopsy, respectively. Yet, in women with overall increased breast density, mammography can miss up to 50% of cancers in the densest breasts and obscure both high contrast microcalcifications, as well as the higher density of some cancers.

Breast magnetic resonance (MR) imaging shows high sensitivity to both ductal carcinoma in situ (DCIS) and invasive cancers due to increased local blood flow and/or permeability to paramagnetic intravenous contrast agents. Limited breast MR work has suggested that the fat-glandular interface (FGI; figure 1) represents a preferable location for up to 94-99% of breast cancers [4,5]. Breast cancer localization at the FGI appears to be a high sensitivity parameter but has not been widely considered. The coronal plane can visualize the circumferential nature of the FGI, which is feasible with 3D acquisitions of breast MR with isotropic pixels, but was not addressed [4,5]. They also described benign solid masses, but cysts, the most common benign masses, were not included. In most women, fat makes up more of the breast than dense tissue, which often leaves part of that coronal circumference

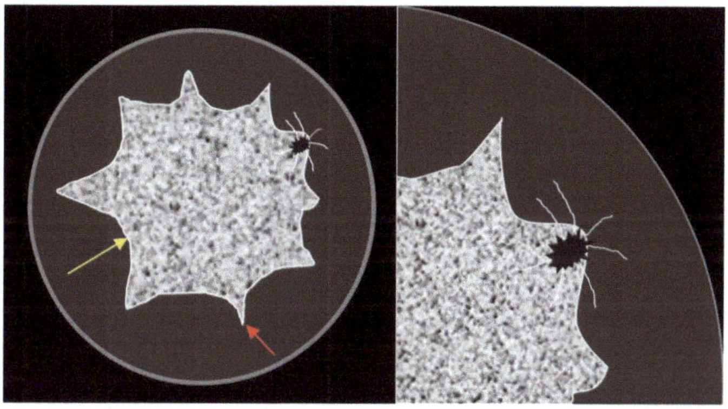

Figure 1: Graphic coronal representations of a UST sound speed (SS) image, highlighting the FGI (yellow arrow): The significantly higher SS of central fibroglandular tissue (speckled) allows good separation from the surrounding fat (darker gray). Red arrow denotes a Cooper's peak. Magnified view (right) shows an irregular cancer (black) at a Cooper's peak with radiating spicules of architectural distortion (white lines).

without parenchyma, leaving only telltale residual fibrous bands (e.g., Cooper‘s ligaments) after involution.

1.2 The Well-Established Cancer Biology at the FGI

Breast cancer initiation and growth have strong associations with adjacent fat cells, or adipocytes, and their fat-secreted hormones, adipokines [6,7]. The FGI also helps define the boundary with the breast‘s subcutaneous adipose layer which may comprise the largest endocrine gland in the body [8]. Adipokines help mediate blood pressure, reproductive function, appetite, glucose homeostasis, angiogenesis and immune function [8]. The link between obesity and the rising incidence of multiple diseases appears to include breast cancer.

When the balance of adipokines tip toward an excessive pro-inflammatory state at the FGI, multiple adipokines, such as leptin, have been implicated in breast cancer initiation via aromatase expression [9,10]. Cancer cell lines and tumor growth become markedly accelerated in the presence of adipocytes [9]. Bidirectional crosstalk between cancer cells and adipocytes, moreover lead to the formation of cancer-associated adipocytes (CAA). These CAA‘s stimulate a phenotypic change to generate fibroblast-like cells known as adipocyte-derived fibroblasts through the secretion of the adipokine, fibronectin, and collagen. Tumor cells cultured with these fibroblasts then demonstrate increased invasive ability. CAA‘s also display overexpression of collagen VI while ECM-related molecules contribute to breast cancer progression. This FGI sequence may also lead to the presence of a dense collagenous stroma, or desmoplastic reaction around breast cancers (i.e., peritumoral region), particularly in estrogen receptor/progesterone receptor (ER/PR) positive tumors.

1.3 UST and the FGI

Ultrasound tomography (UST) development over 40 years [11-36] has accelerated with recent rapid progress of computing capacity and appears well-suited to evaluate the circumferential nature of the FGI in its native coronal plane. A ring array transducer has provided whole breast and focal mass evaluation by combining reflection signal acquisition with quantitative transmission properties of sound speed (SS) and attenuation (ATT). UST offers quantitative multi-parametric evaluation of breast tissues that are scanned and displayed in its highest resolution, coronal plane. Correlation of mammographic breast density to SS distributions of parenchymal tissues throughout the breast have been thoroughly explored [26-36], including improvements in SS resolution [34]. The excellent SS contrast between low SS fat and denser parenchyma/stromal tissues produced even better correlation with MR parenchymal patterns, as well as the volumetric assessment of the entire breast [37]. SS has also been shown to be a stronger BC risk factor than mammographic density [32].

Given the strong biological background for cancer initiation near the FGI, we sought to clarify tissue localization for cancers, solid benign masses and cysts, in relation to their standard anatomic quadrants and tissue components using UST and its innate coronal imaging plane. In a smaller subset, mass evaluations by UST was also able to compare and assess quantitative peritumoral measurements.

2 Methods

For qualitative anatomic and FGI locations, a total of 602 clinical breast masses were noted within 467 individual breasts from 408 patients, as part of a related study of UST for women with dense breasts (SoftVue, Delphinus Medical Technologies Inc., Novi, Michigan; Clinicaltrials.gov – NCT#02977247). This data set included a subset with quantitative region-of-interest (ROI) analyses for the tumor and peri-tumoral regions noted below. Some women had more than one mass in each or both breasts. All masses were biopsy-confirmed by subsequent or prior histology, unless considered as a characteristic cyst by standard HHUS evaluation.

Mass location in coronal UST images was determined by a MQSA certified radiologist with extensive UST imaging experience. The relationship of a mass to the FGI was noted for a all 602 breast masses (181 cancers, 188 fibroadenomas, 173 cysts and 60 other benign findings), confirmed by biopsy, or by hand-held ultrasound for a simple cyst. The 60 other benign findings were excluded from the figures due to their smaller numbers and contained non-mass findings, such as focal fibrosis, micro-cystic changes and/or granulomatous mastitis.

The FGI was defined as the circumferential interface between the peripheral fat and parenchyma [4,5] as determined by examining sound speed (SS) images, which reliably differentiate the much lower SS (i.e., density) of fat [26-36] from the higher SS of fibro-glandular tissues. Mass location was determined from its best visualized coronal level and placed into 3 groups: (i) completely surrounded by dense high SS fibroglandular tissue, (ii) completely surrounded by low SS fat or (iii) partially surrounded by both (i.e., the FGI) [4]. Typically, if a mass showed $\sim$ 1/8 of its circumference abutting fat in its greatest visualized coronal level, it was considered to be at the FGI (figure 2-Fibroadenoma). Conversely, a mass could be considered to abut parenchyma while being surrounded by fat and also labeled at the FGI (figure 2-Cancers). Masses were evaluated only on the coronal plane and therefore not fully evaluated in 3D for FGI location at this time due to lower out of plane resolution of SoftVue imaging (below). Mass location was only augmented by the lower resolution reconstructions in the axial/sagittal reconstructions as needed for additional confirmation of anterior/posterior aspects of the FGI. Masses were also categorized according to standard quadrant positions within the breast based on distance from the nipple and clock position of the finding. Differences in lesion locations were assessed using the chi-squared test.

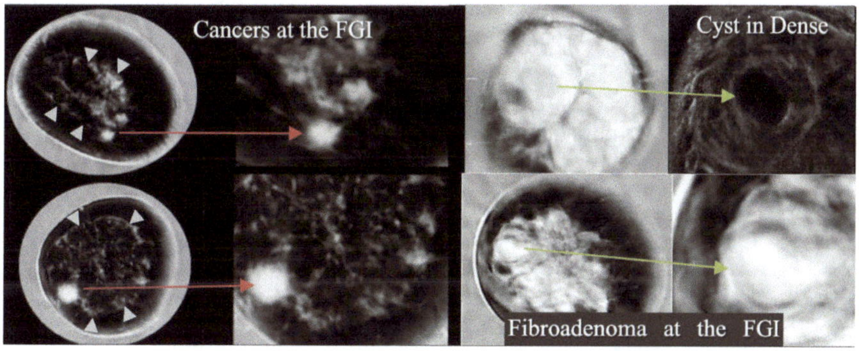

Figure 2: **Cancers, cyst and fibroadenoma:** (Left) Sound speed (SS) images from 2 patients with scattered fibroglandular breast density and high SS cancers at the FGI. Both cancers have irregular margins on the magnified views, being nearly surrounded by fat along the residual fibrous bands of the FGI (arrowheads). Top right SS and magnified reflection view show dominant replacement by extremely dense breast parenchyma (white) with an embedded simple cyst. The well circumscribed 2.0 x 1.3 cm simple cyst has intermediate sound speed, similar to the gray water surrounding the breast. Bottom right images show a patient with heterogeneously dense breast parenchyma and a 1.6 x 1.2 cm fibroadenoma with a lucent halo of lower SS, sometimes also noted with mammography. While mostly embedded in parenchymal tissue, a smooth portion of the fibroadenoma bulges the FGI into the surrounding subcutaneous fat.

For a subset of 296 masses (78 cancers, 105 fibroadenomas, 91 cysts and 24 other benign findings), their boundaries were hand-traced by a radiologist with over 20 years experience as a certified breast imager using MIM software to generate regions of interest (ROI) encompassing the detected masses by UST. Mass margins were defined on the single best visualized/representative image upon a combination of sound speed and reflection image stacks to trace their contours, generating a surface area. Once tumor margins were defined, a peritumoral region was then digitally created to extend 20% of the tumor's diameter beyond all tumor margins, making it proportionate for every tumor. The quantitative parameters of SS and the peritumoral surface area could then be compared with the boundaries inside the traced mass ROI. The SS image was also segmented into two regions corresponding to dense and fatty tissues by using a k-means segmentation method. Combining the segmented SS image with the ROI masks allowed for calculations of the percent density (PD) [the area of dense tissue compared to the total area] for each region as well. Comparisons of mean values between the mass types were performed using an ANOVA analysis. Chi-squared test was used to assess significance of frequency differences.

3 Results

Coronal imaging by UST provided reliable circumferential evaluation of the FGI, allowing a consistent peripheral visual search pattern. Cancers were classified at the FGI in 93.3% (169/181) of the cases, were surrounded by fat in 5.5% (10/181) or parenchyma in 1.1% (2/181)(chi-squared, $p<0.001$) of cases (figure 3). Even when surrounded by fat, cancers were still associated with residual fibrous bands (i.e., Cooper's ligaments) of the atrophied parenchyma at the FGI. For fibroadenomas, 68.6% (129/188) were classified at the FGI while 39.8% (69/173) of cysts were found at the FGI. Moreover, 56.1% (97/173) of cysts and 25.0% (47/188) of fibroadenomas were fully surrounded by dense tissue, significantly more than cancers (1.1%) (chi-squared, $p<0.001$). Few cancers, fibroadenomas or cysts were completely surrounded by fat (i.e., 5.5%, 6.4% and 4.0%, respectively).

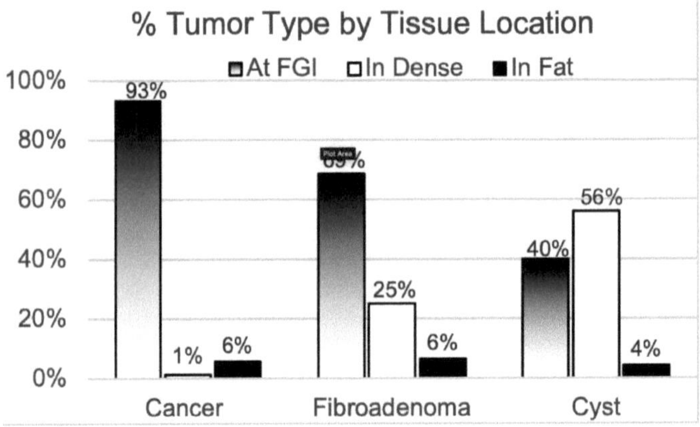

Figure 3: Relative distribution of 3 tissue locations for each tumor type (i.e., 3 locations = 100%). Significant opposing trends are noted for location incidence of Cancer, Fibroadenoma and Cyst at the FGI (shaded bars) versus being surrounded by dense breast parenchyma (white bars). A low percentage of all tumors are surrounded only by fat (black bars).

Figure 4 shows the four-quadrant distribution of cancers, fibroadenomas and cysts. Significantly greater cancer distribution of 42% (76/181) (chi-squared, $p = 0.008$) was noted in the upper outer quadrant compared to cancer rates in the other quadrants, yet a similar greater relative percentage of 34% (63/188) for fibroadenomas and 39.3% (68/173) of cysts were also more common in the upper outer quadrant than the other locations (Chi square $p<0.01$). All tumors were also least commonly located in the lower inner quadrant ($p<0.05$).

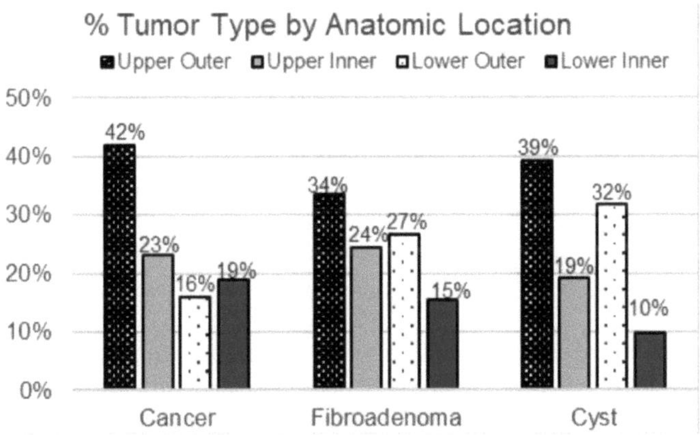

Figure 4: Relative distribution of 4 anatomic breast quadrants for each tumor type (i.e., 4 locations = 100%). All tumors were more commonly located in the upper outer quadrant and have their lowest incidence in the lower inner quadrant, but otherwise showed no mass-specific intra-quadrant differentiation.

Quantitative SS results for surface areas generated for the traced ROIs of the mass showed overlap between masses, but still a significant difference in mean SS (m/sec) for the actual mass ROIs between cancers (1527 m/s, 1521-1533 95%CI) fibroadenomas (1536 m/s, 1531-1540 95% CI) and cysts (1535 m/s, 1529-1540 95% CI)($p = 0.046$). Surrounding peritumoral regions showed significantly lower mean peritumoral SS for cancers (1477 m/s, 1470-1482 95% CI) than fibroadenomas (1496 m/s, 1490-1502 95% CI) as well as lower mean peritumoral SS for fibroadenomas than cysts (1518 m/s, 1512-1523 95% CI)(ANOVA $p<0.001$). The percent density (PD) of the peritumoral region for the different mass types also showed a progressive increase in the mean values when going from cancer to fibroadenoma to cyst (47% to 65% to 84% respectively, ANOVA $p<0.001$). Furthermore, the median peritumoral PD for cysts and cancers are 98.5% and 44.7%, quantitatively indicating that roughly half of all cysts are almost entirely surrounded by dense tissue, while significantly more fat surrounds cancers.

4 Discussion

Our results confirm the common breast imaging dictum of cancers being more common in the upper outer quadrant, but also showed a similar incidence for fibroadenomas and cysts. Similar to prior MRI findings [4,5], UST tissue localization also confirmed cancers had a very strong tendency to be located at the FGI and extend these findings by demonstrating that benign masses have a weaker tendency in that regard. Moreover, quantitative peritumoral SS characteristics of the tumor microenvironment with cancers provides insights to their greater likelihood of being partially surrounded by fat compared to benign masses and support the qualitative findings.

UST is a novel breast imaging technology being explored for its potential for improving breast cancer detection in women with dense breasts (NCT#02977247). The native coronal imaging plane by SoftVue UST may facilitate simpler qualitative detection and localization of cancers, which were much more likely to be found at the FGI (figures 1, 2). Conversely, it appears rare that cancers are entirely surrounded by fibroglandular tissues, or cysts surrounded by fat (i.e., 1.1% and 4.0%, respectively). Even though overall benign masses were significantly more likely to be surrounded by fibroglandular tissue than cancers, only cysts were predominantly surrounded by parenchyma (i.e., 56.1%). These findings are further corroborated by insights provided by the mean peritumoral analyses.

UST separation of breast parenchyma from fat by SS, using segmentation K-means segmentation techniques, have been well established and correlated with breast density by mammogram and parenchymal distribution by MR [26-36]. The mean SS results for mass ROIs and peritumoral regions are better understood when we note that the mean SS for the surrounding water bath is approximately 1500 m/s. Considering that simple cysts contain similar water-like SS or density, smaller cysts appear more often to contain thicker proteinaceous fluid or inspissated debris that can markedly elevate SS.

Data from our peritumoral analyses lend insights to the average tissue composition surrounding masses. Progressively increasing mean SS in the peritumoral tissues of cancers, fibroadenomas and cysts [i.e., 1470-1482 m/s, 1490-1502 m/s and 1512-1523 m/s respectively (ANOVA $p<0.001$)], suggest a similar physiologic progression, from a mix of fatty tissues to dense fibroglandular tissue. For cancers, the 25th percentile to 75th percentile range of mean peritumor SS is 1455 – 1499 m/s, indicating a progression of tissue composition from predominantly fatty to fairly dense, features compatible with tissues found at the FGI. Conversely, the quantitative peritumoral SS range for cysts of 1503 - 1542 m/sec is also compatible with the qualitative assessment of cysts being entirely surrounded by parenchyma a majority of the time. Using the peritumoral PD measurements also supports the qualitative analysis. The 25-75th percentile range for cancer peritumor PD goes from 14% to 75% dense, while for cysts, the same percentile range is 75% to 100% dense. Once again, this shows that the regions surrounding cancers is much more likely to be a mix of dense and fatty tissue while cysts are far more likely to be almost entirely surrounded by only dense tissue.

The preferential tumor location for cancers at, or near, the FGI thus appears to be a highly sensitive finding for cancer origin and growth, but not necessarily specific in this clinical case selection. However, the knowledge of nearly all cancers being near the FGI suggests a reasonable circumferential search pattern that will also require the use of common BI-RADS criteria. In other words, mass margin irregularity noted near the FGI is highly suspicious. Conversely, a smoothly marginated mass near the FGI still requires consideration of fibroadenoma versus more uncommon but well-marginated cancers (e.g., mucinous, medullary, etc.).

Future use as a search tool near the FGI requires comparison with an actual screening series, but the quantitative aspects of UST mass and peritumoral analyses suggest future development with computer-aided diagnosis and detection. Finally, the quantitative tumoral:peritumoral information can be further utilized to better understand the complex biology of adipocytes, adipokines and cancers cells near the FGI.

Numerous weaknesses are inherent in an analysis of a subjective criteria, such as the location of whether a breast tumor resides at the FGI. Rather than a more graded approach [4], we therefore chose a simpler assessment whereby of the extremes of being entirely surrounded by fat or fibroglandular tissue was more straightforward. While the coronal plane is convenient for the circumferential assessment of the FGI, we also acknowledge that further work is needed on the 3D assessment of both anterior and posterior tumor margins near the FGI. As a 3D structure, figure 5 is offered as a graphic rendering to show that the FGI is dispersed throughout the breast volume , but a search can be simplified to an irregularity along the FGI margin.

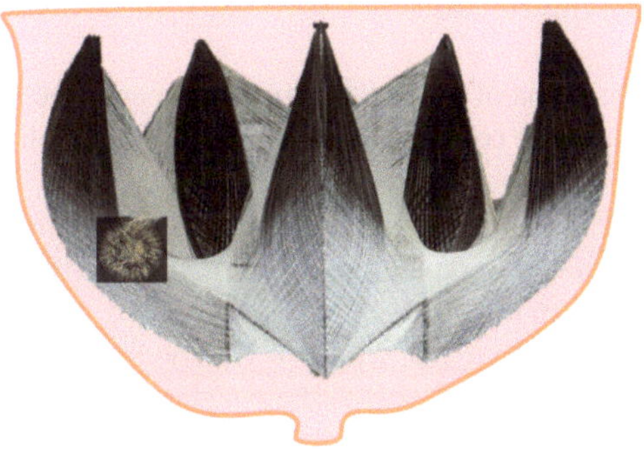

Figure 5: Graphic representation of a pendulus breast (pink background) in what could be postulated as an extremely dense breast with the FGI represented by an inverted architectural dome having a multi-pointed hemispherical mesh (black and gray). The irregularity of a cancer is represented by a burr at one of these points, or Cooper's peak.

The FGI criteria was qualitative but produced significant mass differentiation that prior anatomic quadrant location did not. The peritumoral analyses was also incomplete for all 602 masses, being available for only 298. However, these quantitative analyses of mean SS values required no subjective component and were highly significant and supported the qualitative assessments of the FGI. Further work is also needed for further confirmation with biological and/or molecular correlates.

In summary, both the qualitative and subjective opinion of a radiologist noting cancers' preferential location at the FGI corroborates similar MR findings but emphasizes the coronal UST location as a circumferential view of the FGI. Moreover, quantitative SS results for individual masses and peritumoral regions also corroborates the well-known biological significance of fat being adjacent to cancers and their associated rate of growth. These FGI analyses showed significantly better mass differentiation than anatomic location, which confirmed that all masses are more common in the upper outer quadrant. This study supports its possible integration into clinical practice and future dense breast screening.

5 Conclusion

The coronal view by UST confirmed that cancers are much more likely to be found at the FGI ($p<0.01$) compared to fibroadenomas and cysts. UST is a novel new technology that has the potential for improved breast cancer detection. This study supports increased accuracy of UST imaging as it becomes integrated into clinical practice.

References

[1] Perkins CI, Hotes J, Kohler BA, Howe HL. Association between breast cancer laterality and tumor location, United States, 1994–1998. Cancer Causes Control 2004; CCC 15:637–645.

[2] Lee AH. Why is carcinoma of the breast more frequent in the upper outer quadrant? A case series based on needle core biopsy diagnoses. Breast (Edinburgh, Scotland) 2005; 14:151–152.

[3] Venkatesan A, Chu P, Kerlikowske K, Sickles EA, Smith-Bindman R. Positive predictive value of specific mammographic findings according to reader and patient variables. Radiology 2009; 250(3):648-57.

[4] Kim WH, Li M, Han W, Ryu HS, Moon WK. The spatial relationship of malignant and benign breast lesions with respect to the fat-gland interface on magnetic resonance imaging. Nature Sci Rep. 2016 Dec 14; 6:39085.

[5] Zhu W, Harvey S, Macura KJ, Euhus DM, Artemov D. Invasive breast cancer preferably and predominantly occurs at the interface between fibroglandular and adipose tissue. Clin Breast Cancer 2017; 17(1):e11-e18.

[6] Yamaguchi J, Ohtani H, Nakamura K, Shimokawa I, Kanematsu T. Prognostic impact of marginal adipose tissue invasion in ductal carcinoma of the breast. Am J Clin Pathol. 2008 Sep;130(3):382-8. doi: 10.1309/MX6KKA1UNJ1YG8VN. PMID:1870141.

[7] Wang YY, Attané C, Milhas D, Dirat B, Dauvillier S, Guerard A, Gilhodes J, Lazar I, Alet N, Laurent V, Le Gonidec S, Biard D, Hervé C, Bost F, Ren GS, Bono F, Escourrou G, Prentki M, Nieto L, Valet P, Muller C. Mammary adipocytes stimu-late breast cancer invasion through metabolic remodeling of tumor cells. JCI Insight. 2017 Feb 23; 2(4):e87489. doi: 10.1172/jci.insight.87489. PMID:28239646.

[8] Mancuso P. The role of adipokines in chronic inflammation. Immunotargets Ther. 2016; 5: 47–56.

[9] Dirat B, Bochet L, Dabek M, Daviaud D, Dauvillier S, Majed B, Wang YY, Meulle A, Salles B, Le Gonidec S, Garrido I, Escourrou G, Valet P, Muller C. Cancer-associated adipocytes exhibit an activated phenotype and contribute to breast cancer invasion. Cancer Res. 2011; 71(7):2455-65.

[10] Bochet L, Lehuédé C, Dauvillier S, Wang YY, Dirat B, Laurent V, Dray C, Guiet R, Maridonneau-Parini I, Le Gonidec S, Couderc B, Escourrou G, Valet P, Muller C.

Adipocyte-derived fibroblasts promote tumor progression and contribute to the desmoplastic reaction in breast cancer. Cancer Res 2013; 73(18):5657-68.
[11] S. Johnson, D.T. Borup, J.W. Wiskin, M. Berggren, B. Hanover, F. Setinsek, et al. From Laboratory To Clinical Trials: An Odyssey Of Ultrasound Inverse Scattering Imaging For Breast Cancer Diagnosis. The Journal of the Acoustical Society of Ame-rica. 120: p. 3023 (2006).
[12] S.A. Johnson and M.L. Tracy ML. Inverse Scattering Solutions By a Sinc Basis, Multiple Source, Moment Method. Part I: Theory, Ultrasonic Imaging, 5:361-375, (1983).
[13] J.S. Schreiman, J.J. Gisvold, J.F. Greenleaf, R.C. Bahn. Ultrasound Transmission Computed Tomography Of The Breast. Radiology; 150:523-30, (1984).
[14] G.H. Glover. Computerized Time-Of-Flight Ultrasonic Tomography For Breast Examination. Ultrasound In Medicine & Biology. 3(2-3):117-27, (1977).
[15] F.A. Natterer. Propagation Backpropagation Method for Ultrasound Tomography, Inverse problems; 11:1225-1232, (1995).
[16] P.L. Carson, C.R. Meyer, A.L. Scherzinger, T.V. Oughton. Breast Imaging In Coronal Planes With Simultaneous Pulse Echo And Transmission Ultrasound. Sci-ence. 214(4525):1141-3, (1981).
[17] M. P. Andre, H.S. Janee, P. J. Martin, G. P. Otto, B. A. Spivey, D. A. Palmer, High-Speed Data Acquisition In A Diffraction Tomography System Employing Large-Scale Toroidal Arrays, International Journal of Imaging Systems and Technol-ogy. 8:137-147, (1997).
[18] S.A. Johnson, D. T. Borup, J.W. Wiskin, F. Natterer, F. Wuebbling, Y. Zhang, C. Olsen Apparatus and Method for Imaging with Wavefields using Inverse Scattering Techniques. United States Patent 6,005,916 (1999).
[19] V.Z Marmarelis, T. Kim, R.E. Shehada, Proceedings of the SPIE: Medical Imaging 2003; San Diego, California; Feb. 23-28, 2002. Ultrasonic Imaging and Signal Processing – Paper 5035-6.
[20] D.L. Liu and R. C. Waag, Propagation and backpropagation for ultrasonic wavefront design, IEEE Trans. on Ultras. Ferro. and Freq. Contr. 44(1):1-13 (1997).
[21] H. Gemmeke and N. Ruiter. 3D Ultrasound Computer Tomography for Medical Imaging. Nuclear instruments and methods in Physics Research Section A: Accelera-tors, Spectrometers, Detectors and Associated Equipment, 580: 1057-1065, (2007).
[22] N. Ruiter, G. Göbel, L. Berger, M. Zapf and H. Gemmeke Realization of an Opti-mized 3D USCT, Proc. SPIE 7968, 796805 (2011)
[23] N. Duric, P.J. Littrup, L. Poulo, A. Babkin, R. Pevzner, E. Holsapple, O. Rama, and C. Glide. Detection of Breast Cancer With Ultrasound Tomography: First Re-sults with the Computed Ultrasound Risk Evaluation (CURE) Prototype. Med Phys. 34: 773-85, (2007).
[24] C. Li, G.S. Sandhu, M. Boone, N. Duric, P. Littrup, M. Sak, and K. Bergman. Breast Tissue Characterization with Sound Speed and Tissue Stiffness. In Proceed-ings of the International Workshop on Medical Ultrasound Tomography: 1.-3. Nov. 2017, Speyer, Germany (p. 217). KIT Scientific Publishing (2018).

[25] B. Ranger, P.J. Littrup, N. Duric, P. Chandiwala-Mody, C. Li, S. Schmidt, J. Lupinacci. Breast Ultrasound Tomography versus MRI For Clinical Display of Anatomy and Tumor Rendering: Preliminary Results. AJR Am J Roentgenol; 198:233-9, (2012).

[26] C. Glide, N. Duric, P. Littrup: Novel Approach to Evaluating Breast Density Utiliz-ing Ultrasound Tomography. Med Phys, 34:744-753, (2007).

[27] C.K. Glide-Hurst, N. Duric, P. Littrup: Volumetric breast density evaluation from ultrasound tomography images. Med Phys , 35:3988-3997, (2008).

[28] N. Duric, N. Boyd, P. Littrup, M. Sak, L. Myc, C. Li, E. West, S. Minkin, L. Martin, M. Yaffe, S. Schmid, M. Faiz, J. Shen, O. Melnichouk, Q. Li, and T. Alb-recht. Breast Density Measurements with Ultrasound Tomography: A Comparison with Film and Digital Mammography. Med Phys; 40(1): 013501, (2013).

[29] N. Duric, M. Sak, S. Fan, R.M. Pfeiffer, P.J. Littrup, M.S. Simon, D.H. Gorski, H. Ali, K.S. Purrington, R.F. Brem, M.E. Sherman, G. Gierach. Using Whole Breast Ultrasound Tomography to Improve Breast Cancer Risk Assessment: A Novel Risk Factor Based on the Quantitative Tissue Property of Sound Speed. Journal of Clini-cal Medicine. 9(2): E367. doi: 10.3390/jcm9020367, (2020).

[30] Z.G. Khodr, M.A Sak, R.M. Pfeiffer, N. Duric, P. Littrup, L. Bey-Knight, H. Ali, P. Vallieres, M.E. Sherman, and G.L. Gierach. Determinants of The Reliability of Ultrasound Tomography Sound Speed Estimates as A Surrogate for Volumetric Breast Density. Medical Physics; 42(10), pp.5671-5678, (2015).

[31] G.Y. Sandhu, C. Li, O. Roy, S. Schmidt and N. Duric. Frequency Domain Ultrasound Waveform Tomography: Breast Imaging Using A Ring Transducer. Physics in Medicine & Biology; 60(14), p.5381 (2015).

[32] M. Sak, N. Duric, P. Littrup, L. Bey-Knight, H. Ali, P. Vallieres, M.E. Sherman, and G.L. Gierach. Using Speed of Sound Imaging to Characterize Breast Density. Ultrasound Med Biol, 43(1): 91-103 (2017).

[33] J. Wiskin, B. Malik, R. Natesan, M. Lenox. Quantitative Assessment of Breast Density Using Transmission Ultrasound Tomography. Med Phys. 46(6):2610-20, (2019).

[34] E.A.M. O'Flynn, J. Fromageau, A.E. Ledger, A. Messa, A. D'Aquino, M.J. Schoemaker, M. Schmidt, N. Duric, A.J. Swerdlow, and J.C. Bamber. Ultrasound Tomography Evaluation of Breast Density: A Comparison with Non-Contrast Magnetic Resonance Imaging. Invest Radiol 52(6): 343-348 (2017).

[35] P.J. Littrup, N. Duric, C. Li, M. Sak, G. Sandhu, K. Bergman, et al. Current challenges in breast screening and diagnosis: From molecules to peritumoral regions and radiomics - The emerging imaging of whole breast stiffness. Proceedings of the International Workshop on Medical Ultrasound Tomography: 1.-3., Speyer, Germany (2017).

[36] N. Duric, P. Littrup, C. Li, O. Roy, S. Schmidt, J. Seamans, A. Wallen, and L. Bey-Knight, L., 2015, March. Whole breast tissue characterization with ultrasound tomography. In Medical Imaging: Ultrasonic Imaging and Tomography, 9419: 94190G (2015). International Society for Optics and Photonics.

[37] E.A.M. O'Flynn, J. Fromageau, A.E. Ledger, A. Messa, A. D'Aquino, M.J. Schoemaker, M. Schmidt, N. Duric, A.J. Swerdlow, and J.C. Bamber. Ultrasound tomography evaluation of breast density: A comparison with non-contrast magnetic resonance imaging. Invest Radiol 52(6): 343-348 (2017).

[38] P.J. Littrup, R.A. Kane, C.R. Williams, T.K. Egglin, F. Lee, S. Torp-Pedersen, P.A. Church. Determination of prostate volume with transrectal us for cancer screen-ing. Part I: Comparison with prostate specific antigen assays. Radiology; 178:537-542 (1991).

[39] P.J. Littrup, C.R. Williams, T.K. Egglin and R.A. Kane. Determination of prostate volume with transrectal us for cancer screening. Part II: The accuracy of in vitro and in vivo techniques. Radiology 1991; 179:49-53 (1991).

Methods

USCT data challenge 2019

C. Boehm[1], T. Hopp[2], and N. Ruiter[2]

[1] *ETH Zurich, Switzerland, E-mail: christian.boehm@erdw.ethz.ch*
[2] *Karlsruhe Institute of Technology, Germany*

Abstract

Recent years have witnessed the active development of scanning systems and reconstruction algorithms for ultrasound computed tomography (USCT) with applications to breast imaging for early cancer detection. Despite these advances in hardware and software development, we encounter the need for reference data to develop, test and compare different imaging methods. With the goals of encouraging scientific exchange and collaborations, providing benchmarks of reconstruction algorithms, and standardizing USCT data formats, we have released open-source data sets of simulated waveforms that mimic measurements of a USCT scanning aperture using numerical breast phantoms. This is part of ongoing efforts centered around the USCT platform for data exchange and collaboration [1, 2].

Keywords: ultrasound computer tomography, numerical modelling, data challenge

1 Introduction

Within the past decade, substantial progress in the development and design of USCT systems has been made, and several systems have readily provided first clinical results [3, 4, 5]. Simultaneously, a wide range of image reconstruction methods using both transmission and reflection data has been developed. Overviews of different modelling and inversion techniques can be found in [6] and [7]. One of the main challenges is to find good trade-offs between the quality of the image, on the one hand, and the time-to-solution, on the other hand. Popular reconstruction methods include ray-based time-of-flight inversion [8, 9, 10], reflection imaging using synthetic aperture focusing techniques [11] and reverse time migration [12], Born modelling [13, 14], as well as waveform tomography and full-waveform inversion in time and frequency domain [15, 16, 17, 18, 19, 20, 21, 22].

While this list is far from being complete, the variety of imaging and reconstruction algorithms readily stresses the need for suitable reference data to benchmark and compare different methods. To meet the demand for freely available data sets within the USCT research

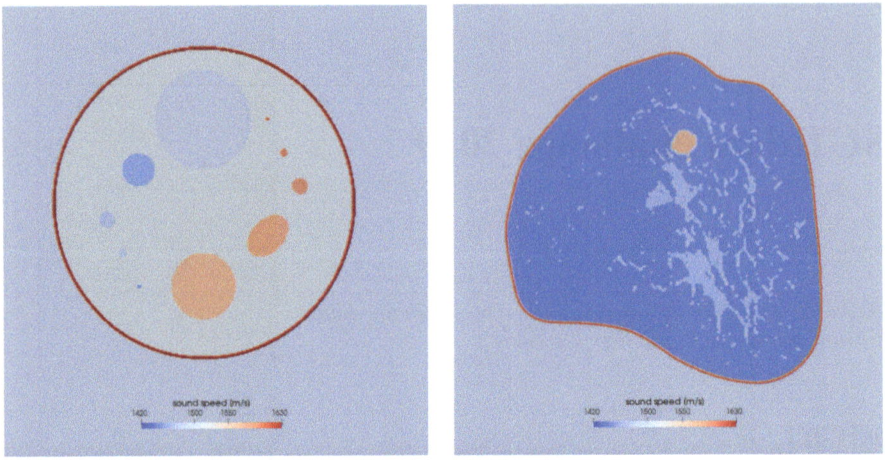

Figure 1: Sound speed map of the analytical **(left)** and the segmented breast phantom **(right)** used in the simulations.

community, a digital platform hosting several data sets has been initiated. In the initial phase this included raw data donated by three different USCT systems [1, 2]. In the course of this new data challenge, we added simulated A-scans using numerical phantoms that enable researchers to compare and benchmark different reconstruction algorithms with a ground truth.

The long term goals of this initiative are (1) to build up an open-source reference USCT data base that is freely available for the entire community, (2) to enable reproducible comparisons of image reconstruction algorithms and USCT systems, (3) to establish user friendly interfaces, standards and data formats between different USCT apertures, reconstruction algorithms, software and data formats, and (4) to identify properties of systems, experimental setups, data, and algorithms towards optimal images.

In the rest of the paper we briefly summarize the specification of the numerical phantoms used in this data challenge, comment on the scanning aperture and simulation setup, and propose a flexible data format for USCT scans.

2 Numerical phantoms

This data challenge comprises waveform data for two different numerical phantoms. Figure 1 shows the sound speed map of both phantoms. In addition, data generated in a homogeneous medium mimicking an empty water tank with a constant sound speed of 1500 m/s and a density of 1000 kg / m^3 is provided for calibration purposes.

Ellipse	**Center** (x, y) [mm]	**Axes** (x, y) [mm]	θ [°]	c [m/s]	ρ [kg / m^3]
a	(0, 0)	(47, 47)	0	1650	1150
b	(0, 0)	(46, 46)	0	1510	1040
c	(0, -25)	(10, 10)	0	1560	1070
d	(0, 25)	(15, 15)	0	1500	990
e	(20, -10)	(5, 7.5)	45	1570	1080
f	(30, 5)	(2.5, 2.5)	0	1580	1090
g	(20, 25)	(0.625, 0.625)	0	1600	1150
h	(-20, 10)	(5, 5)	0	1470	980
i	(-30, -5)	(2.5, 2.5)	0	1490	1000
j	(-25, -15)	(1.25, 1.25)	0	1490	1100
j	(-20, -25)	(0.625, 0.625)	0	1450	960

Table 1: The geometric parameters used for creating the analytic phantom as well as the values for sound speed c and density ρ. The order is important and later added shapes overwrite the previous values.

Tissue	c [m/s]	ρ [kg / m^3]
water	1500	1000
fat	1450	980
glandular tissue	1480	1020
skin	1600	1060
muscle	1580	1080
irregular cancer	1550	1070
lobular cancer	1560	1010
cyst	1500	1000
fibroadenoma	1530	1080

Table 2: Values of sound speed c and density ρ for different tissues labelled in the segmented phantom.

2.1 Analytic shapes

The first breast phantom is a superposition of spheres and ellipses with different sizes and constant material properties, and has a diameter of 94 mm. Table 1 provides the full description of all shapes.

2.2 Segmented MRI data

The second phantom is a 2D coronal slice of segmented MRI data of a human breast. The discrete values assigned to the different tissue labels are given in Table 2. As exact tissue parameters are still controversially discussed, we empirically selected values which are in the range of the values reported across the literature [23]. The resolution of the original MRI is fairly coarse with a voxel size of 0.68359 mm and 311 $\times$ 316 pixels in the coronal plane. This yields an irregular pattern, which is primarily visible in the skin. At high frequencies, this stair-casing effect can lead to significant distortion and spurious reflections as shown in Figure 2. To mitigate this effect, we smooth the skin layer in a preprocessing step.

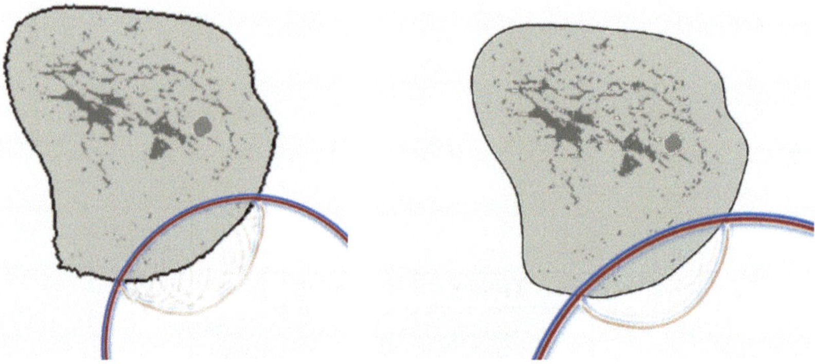

Figure 2: Snapshots of the pressure field. Using the raw MRI data to obtain the speed map leads to signifcant distortion at the water-skin interface **(left)**. This unphysical effect can be mitigated by smoothing the sound speed model in a preprocessing step **(right)**.

3 Aperture and simulated data

We simulate the time-domain acoustic wave equation parameterized by a heterogeneous density ρ and sound speed c:

$$\frac{1}{\rho(\mathbf{x})c(\mathbf{x})^2}\partial_{tt}\phi(\mathbf{x},t) - \nabla \cdot \frac{1}{\rho(\mathbf{x})}\nabla\phi(\mathbf{x},t) = s(t)\delta(\mathbf{x}-\mathbf{x}_s), \tag{1}$$

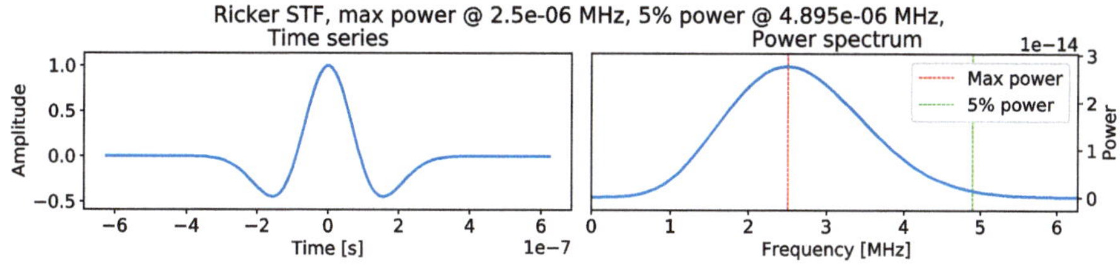

Figure 3: Source wavelet in time domain and its power spectral density.

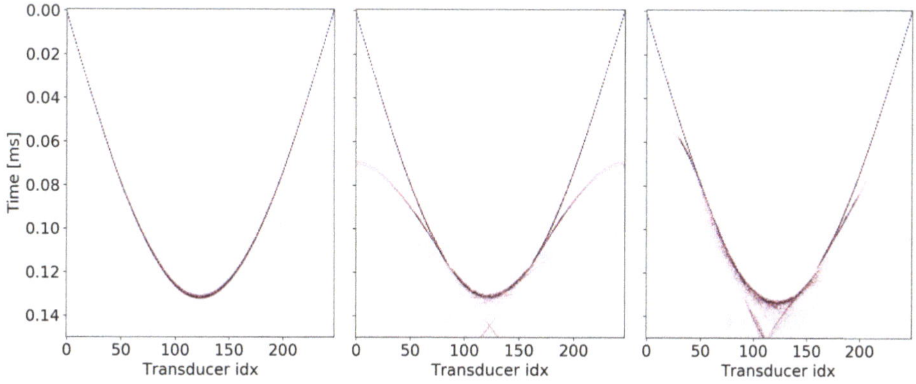

Figure 4: Signal gathers for synthetic waveforms computed in water **(left)**, the analyitc phantom **(middle)**, and the segmented phantom **(right)**. The colorscale is centered around zero and clips large amplitudes to highlight reflected and refracted waveforms after the first arrival.

with homogenuous initial conditions

$$\phi(\mathbf{x},0) = \partial_t \phi(\mathbf{x},0) = 0, \tag{2}$$

where ϕ denotes the time- and space-dependent wavefield and s is the source-time function of an omni-directional point source emitting at location $\mathbf{x}_s$. The source time function is a Ricker wavelet, centered around time zero. The dominant frequency for the 2D data is 2.5 MHz, and the frequency cutoff of 5% of the power spectrum is close to 5 MHz, cf. Fig. 3. The 2D simulations use an aperture with 249 transducers aligned on a ring with a radius of 99 mm. The simulation domain uses a box extended along both coordinate axis and with absorbing boundary conditions imposed on all side walls. The duration of the simulation spans 0.150635 ms, and the data are sampled with a time increment of 9.4 ns. Fig. 4 shows collected A-scans for a single emitter in a homogenuous medium and both numerical phantoms. Reflections from the skin and the stronger sounds speed anomalies can be distinguished in the data of the analytical phantom. This is not the case for the segmented phantom All simulations were carried out using the spectral-element method [24].

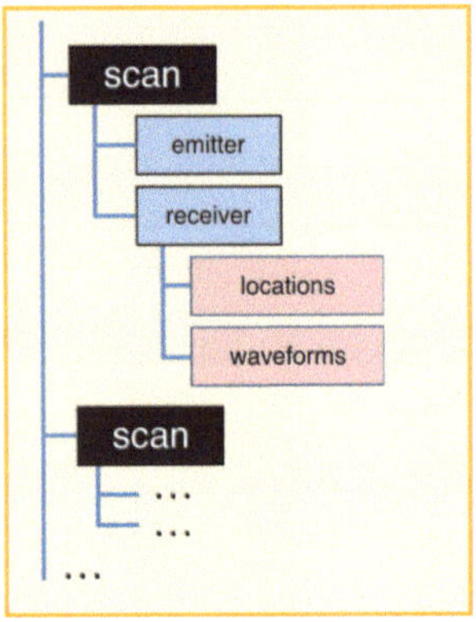


```
HDF5 "usct_data.h5" {
GROUP "/" {
   GROUP "scans" {
      GROUP "scan_0000" {
         GROUP "emitter" {
            DATASET "location" {
               DATATYPE  H5T_IEEE_F64LE
               DATASPACE  SIMPLE { ( dim ) / ( dim ) }
            }
         }
         GROUP "receiver" {
            DATASET "locations" {
               DATATYPE  H5T_IEEE_F64LE
               DATASPACE  SIMPLE { ( n_tdx, dim ) / ( n_tdx, dim ) }
            }
            DATASET "waveforms" {
               DATATYPE  H5T_IEEE_F32LE
               DATASPACE  SIMPLE { ( n_t, n_tdx ) / ( n_t, n_tdx ) }
               ATTRIBUTE "sampling_rate_in_hertz" {
                  DATATYPE  H5T_IEEE_F64LE
                  DATASPACE  SCALAR
               }
               ATTRIBUTE "start_time_in_seconds" {
                  DATATYPE  H5T_IEEE_F64LE
                  DATASPACE  SCALAR
               }
            }
         }
      }
   }
}
}
```

Figure 5: Sketch of the container structure within **(left)** and how this translates into groups and datasets in the hdf5 file **(right)**. Note that additional meta data can easily be stored as attributes or auxiliary datasets. Data types and data set dimensions are dynamic.

4 Data format

To the best of our knowledge, no standardized file format to easily store, process and transfer USCT raw data has emerged from the community yet. As part of releasing data for this challenge, we propose a flexible data format based on HDF5 [25], a widespread, open-source library for storing scientific data. HDF5 has built-in support for compression, chunking, check summing and variable floating point precision. Internally, the file is organized in a container-like structure similar to a filesystem tree. This permits storing waveform data and meta information together in a single file. Furthermore, data from multiple scans can easily be combined or split into several files.

Figure 5 provides a high-level view on the proposed container structure. All information can be accessed individually, which permits lazy loading and batch-wise operations if memory is scarce. Wrappers for reading and writing file contents are currently available for Python, Matlab, C++. We also provide a validation tool to ensure compliance with the file format definition.

5 Conclusions

The aim of this work is to provide freely available simulated USCT data to test and benchmark different imaging algorithms. These simulated data of numerical phantoms follow the previous USCT data challenge that provided measurements from different scanning systems.

The data and a small collection of software tools are available open-source, and we hope to also include inversion and imaging algorithms in the future. We will continue to maintain and extend this platform, and we welcome active participation and feedback. This will contribute to the overarching goals of this initiative, which are to facilitate research on early breast cancer detection, improve imaging algorithms and scanning apertures, and to foster scientific collaborations.

Acknowledgement

The authors would like to thank Òscar Calderón Agudo, Antonio Stanziola, Lluís Guasch, Naiara Korta Martiartu, and Ines Ulrich for their active participation. Furthermore, we thank Ulas Taskin, Joaquin L. Herraiz, Lion Krischer and Neb Duric for their advice and support in initating this data challenge. This work was supported by a grant from the Swiss National Supercomputing Centre (CSCS) under project ID s922.

References

[1] N. V. Ruiter, M. Zapf, T. Hopp, H. Gemmeke, and K. W. A. van Dongen, "USCT data challenge," in *Medical Imaging 2017: Ultrasonic Imaging and Tomography* (N. Duric and B. Heyde, eds.), vol. 10139, pp. 412 – 419, International Society for Optics and Photonics, SPIE, 2017.

[2] N. V. Ruiter, M. Zapf, T. Hopp, H. Gemmeke, K. W. A. van Dongen, J. Camacho, J. L. Herraiz, M. P. Liva, and J. M. Udías, "USCT reference data base: conclusions from the first SPIE USCT data challenge and future directions," in *Medical Imaging 2018: Ultrasonic Imaging and Tomography* (N. Duric and B. C. Byram, eds.), vol. 10580, pp. 170 – 176, International Society for Optics and Photonics, SPIE, 2018.

[3] M. André, J. Wiskin, and D. Borup, "Clinical results with ultrasound computed tomography of the breast," in *Quantitative Ultrasound in Soft Tissues* (J. Mamou and M. L. Oelze, eds.), pp. 395–432, Dordrecht: Springer Netherlands, 2013.

[4] N. Duric, P. Littrup, C. Li, O. Roy, S. Schmidt, X. Cheng, J. Seamans, A. Wallen, and L. Bey-Knight, "Breast imaging with SoftVue: initial clinical evaluation," in *Medical Imaging 2014: Ultrasonic Imaging and Tomography* (J. G. Bosch and M. M. Doyley, eds.), vol. 9040, pp. 208 – 215, International Society for Optics and Photonics, SPIE, 2014.

[5] H. Gemmeke, T. Hopp, M. Zapf, C. Kaiser, and N. V. Ruiter, "3d ultrasound computer tomography: Hardware setup, reconstruction methods and first clinical results," *Nuclear Instruments and Methods in Physics Research Section A: Accelerators, Spectrometers, Detectors and Associated Equipment*, vol. 873, pp. 59 – 65, 2017. Imaging 2016.

[6] R. J. Lavarello and A. J. Hesford, "Methods for forward and inverse scattering in ultrasound tomography," in *Quantitative Ultrasound in Soft Tissues* (J. Mamou and M. L. Oelze, eds.), pp. 345–394, Dordrecht: Springer Netherlands, 2013.
[7] N. Ozmen, R. Dapp, M. Zapf, H. Gemmeke, N. V. Ruiter, and K. W. A. van Dongen, "Comparing different ultrasound imaging methods for breast cancer detection," *IEEE Transactions on Ultrasonics, Ferroelectrics, and Frequency Control*, vol. 62, pp. 637–646, April 2015.
[8] S. Li, M. Jackowski, D. P. Dione, T. Varslot, L. H. Staib, and K. Mueller, "Refraction corrected transmission ultrasound computed tomography for application in breast imaging," *Medical Physics*, vol. 37, no. 5, pp. 2233–2246, 2010.
[9] M. Pérez-Liva, J. L. Herraiz, L. Medina-Valdés, J. Camacho, C. Fritsch, B. E. Treeby, and J. M. Udías, "Regularization of image reconstruction in ultrasound computed tomography," in *2015 IEEE Nuclear Science Symposium and Medical Imaging Conference (NSS/MIC)*, pp. 1–3, Oct 2015.
[10] T. Hopp, M. Zapf, H. Gemmeke, and N. V. Ruiter, "Experimental evaluation of straight ray and bent ray phase aberration correction for USCT SAFT imaging," in *Medical Imaging 2018: Ultrasonic Imaging and Tomography* (N. Duric and B. C. Byram, eds.), vol. 10580, pp. 136 – 148, International Society for Optics and Photonics, SPIE, 2018.
[11] T. Hopp, F. Zuch, M. Zapf, H. Gemmeke, and N. V. Ruiter, "Experimental analysis of ray-based sound speed reconstruction algorithms for phase aberration corrected USCT SAFT imaging," in *Medical Imaging 2019: Ultrasonic Imaging and Tomography* (B. C. Byram and N. V. Ruiter, eds.), vol. 10955, pp. 128 – 138, International Society for Optics and Photonics, SPIE, 2019.
[12] O. Roy, M. A. H. Zuberi, R. G. Pratt, and N. Duric, "Ultrasound breast imaging using frequency domain reverse time migration," in *Medical Imaging 2016: Ultrasonic Imaging and Tomography* (N. Duric and B. Heyde, eds.), vol. 9790, pp. 84 – 92, International Society for Optics and Photonics, SPIE, 2016.
[13] J. Wiskin, D. Borup, S. Johnson, and M. Berggren, "Non-linear inverse scattering: High resolution quantitative breast tissue tomography," *The Journal of the Acoustical Society of America*, vol. 131, no. 5, pp. 3802–3813, 2012.
[14] N. K. Martiartu, C. Boehm, and A. Fichtner, "3d wave-equation-based finite-frequency tomography for ultrasound computed tomography," 2019.
[15] R. G. Pratt, L. Huang, N. Duric, and P. Littrup, "Sound-speed and attenuation imaging of breast tissue using waveform tomography of transmission ultrasound data," *Proc. SPIE*, vol. 6510, p. 65104S, 2007.
[16] G. Sandhu, C. Li, O. Roy, S. Schmidt, and N. Duric, "Frequency domain ultrasound waveform tomography: breast imaging using a ring transducer," *Physics in medicine and biology*, vol. 60, no. 14, p. 5381, 2015.
[17] A. Goncharsky, S. Y. Romanov, and S. Y. Seryozhnikov, "A computer simulation study of soft tissue characterization using low-frequancy ultrasonic tomography," *Ultrasonics*, vol. 67, pp. 136–150, 2016.

[18] T. P. Matthews, K. Wang, C. Li, N. Duric, and M. A. Anastasio, "Regularized dual averaging image reconstruction for full-wave ultrasound computed tomography," *IEEE Transactions on Ultrasonics, Ferroelectrics, and Frequency Control*, vol. 64, no. 5, pp. 811–825, 2017.

[19] M. Perez-Liva, J. L. Herraiz, J. M. Udias, E. Miller, B. T. Cox, and B. E. Treeby, "Time domain reconstruction of sound speed and attenuation in ultrasound computed tomography using full wave inversion," *The Journal of the Acoustical Society of America*, vol. 141, no. 3, pp. 1595–1604, 2017.

[20] A. Ihrig and G. Schmitz, "Accelerating nonlinear speed of sound reconstructions using a randomized block kaczmarz algorithm," in *2018 IEEE International Ultrasonics Symposium (IUS)*, pp. 1–9, Oct 2018.

[21] C. Boehm, N. K. Martiartu, N. Vinard, I. J. Balic, and A. Fichtner, "Time-domain spectral-element ultrasound waveform tomography using a stochastic quasi-Newton method," in *Medical Imaging 2018: Ultrasonic Imaging and Tomography* (N. Duric and B. C. Byram, eds.), vol. 10580, pp. 92 – 100, International Society for Optics and Photonics, SPIE, 2018.

[22] O. C. Agudo, L. Guasch, P. Huthwaite, and M. Warner, "3D imaging of the breast using full-waveform inversion," in *Proceedings of the International Workshop on Medical Ultrasound Tomography: 1.-3. Nov. 2017*, p. 99, KIT Scientific Publishing, 2018.

[23] P.A. Hasgall, F. Di Gennaro, C. Baumgartner, E. Neufeld, B. Lloyd, M.C. Gosselin, D. Payne, A. Klingenböck, N. Kuster, "IT'IS Database for thermal and electromagnetic parameters of biological tissues," Version 4.0," 2018.

[24] M. Afanasiev, C. Boehm, M. van Driel, L. Krischer, M. Rietmann, D. A. May, M. G. Knepley, and A. Fichtner, "Modular and flexible spectral-element waveform modelling in two and three dimensions," *Geophysical Journal International*, vol. 216, no. 3, pp. 1675–1692, 2019.

[25] The HDF Group, "Hierarchical Data Format, version 5," 1997-2017. http://www.hdfgroup.org/HDF5/.

Multi-Parameter Inversion

U. Taskin[1] and K.W.A. van Dongen[1]

[1] *Delft University of Technology, Delft, the Netherlands, E-mail: u.taskin@tudelft.nl*

Abstract

Ultrasound computed tomography (USCT) is gaining interest as a scanning modality to detect breast cancer in woman. To reconstruct the acoustic tissue parameters, full-wave inversion methods are applied on the measured data. However, in many works, only the speed of sound profile of the medium is reconstructed and the mass density is assumed to be constant. In this work, we aim to reconstruct simultaneously the compressibility and mass density from the measured pressure field. We use a full-vectorial contrast source inversion (CSI) procedure where measurements of both the velocity field and the pressure field are required to reconstruct the mass density and compressibility of the medium. In practice, only the pressure field is measured. To reconstruct the velocity field from the pressure field, we developed a method based on Hankel function decomposition of the pressure field. After the velocity field is reconstructed, the full-wave inverse problem is solved for the unknown compressibility and mass density.

Keywords: full-wave inversion, multi-parameter inversion, breast ultrasound, density

1 Introduction

Breast cancer is the most frequently diagnosed cancer type among woman. Mammography is the main screening modality for breast cancer detection. However, mammography has a low sensitivity with dense breasts as both the cancer and the breast tissue show up white on a mammogram. Ultrasound has shown to be useful for this case and could be used as a complementary screening modality [1]. In addition, ultrasound is cheap, safe and patient-friendly. Recent works with ultrasound focused on building a water bath scanning system for breast cancer detection [2, 3]. With this kind of setup, both tranmission and reflection measurements are obtained. This leads to high resolution quantitative images especially when (full-wave) inversion methods are used [4, 5, 6].

One commonly made approximation with full-wave inversion is that the mass density is assumed to be constant and therefore only speed of sound or compressibility of the medium is reconstructed. This is shown to be a good approximation for breast cancer applications especially as inverting for both mass density and compressibility is known to be problematic with the pressure field only [7]. However, multi-parameter inversion becomes more robust with the availability of the velocity field [8]. Furthermore, accurate multi-parameter reconstruction provides better tissue characterization. Unfortunately, in a water bath scanning setup only the pressure field is measured.

In this work, we propose a method to compute the velocity field from the pressure field measurement. The method is based on Hankel decomposition of the measured pressure field [9]. After decomposing the pressure field, we are able to compute the velocity field. Finally, we use the measured pressure field together with the computed velocity field in our full-wave inversion method to accurately reconstruct mass density and compressibility profiles of the medium.

2 Theory

2.1 Multi-parameter inversion

The aim of inversion is to reconstruct the acoustic medium parameters $\mathbf{m}$ from the measured ultrasound signals. To this end, the misfit between the measured field $\mathbf{d}^{\text{meas}}(\mathbf{m})$ and computed field $\mathbf{d}^{\text{comp}}(\mathbf{m})$ at the measurement domain $\mathbb{S}$ is minimized, i.e.,

$$\text{Err} = ||\mathbf{d}^{\text{meas}}(\mathbf{m}) - \mathbf{d}^{\text{comp}}(\mathbf{m})||^2_{\mathbb{S}}. \tag{1}$$

With traditional single-parameter inversion, $\mathbf{m}$ represents only the speed of sound. For that case, measurement of the pressure field is enough to solve the inverse problem. With multi-parameter inversion, $\mathbf{m}$ represents more than one single parameter, e.g., compressibility, density, and/or attenuation. In that case, the complexity of the problem increases and some additional regularization is needed. In a previous work, it is shown that using the velocity field together with the pressure field allows one to reconstruct both compressibility and density [8]. The lack of velocity field measurement is addressed in the next section.

2.2 Velocity field computation

The outward propagating pressure field $p(r,\phi)$ measured on an arbitrary closed curve can be described with Hankel functions as follows

$$p(r,\phi) = \sum_{n=-N}^{N} c_n H_n^{(1)}\left(\frac{\omega}{c_0} r\right) e^{jn\phi}, \tag{2}$$

where c_n represents complex coefficients and $H_n^{(1)}$ is the Hankel function. The unknown coefficients c_n can be found by matching the computed pressure field to the measurement.

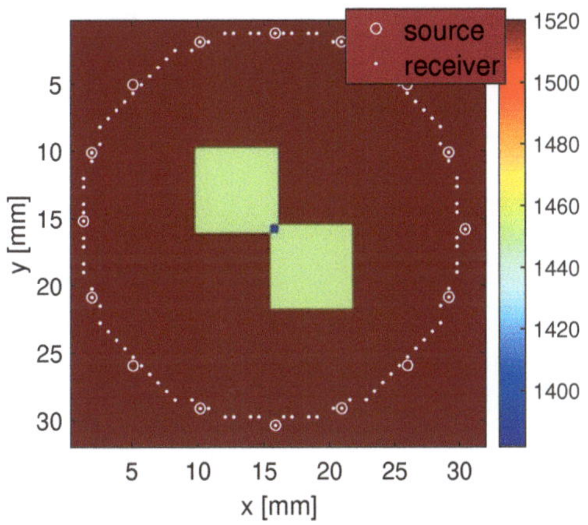

Figure 1: Synthetic experiment setup. Colorbar indicates speed of sound (m/s) values.

After constructing the coefficients c_n, the pressure field can be computed at any point in the homogeneous embedding. Next, the velocity field is computed by taking the gradient of the reconstructed pressure field.

3 Results

3.1 Configuration

A simple synthetic example is used for testing the velocity field computations and multi-parameter inversions. The object of interest is chosen to be two squares with different compressibility and density values. There are 16 sources and 100 receivers equally distributed on a circle that encloses the two objects, see Figure 1. The signal content of the source excitation is given in Figure 2. First, the full-wave forward problem is solved to obtain both pressure and velocity field measurements at the receiver locations.

3.2 Velocity field comparison

Since the original velocity field is known (from the full-forward solution), a comparison between the original and the reconstructed velocity field can be made. The reconstructed velocity field is obtained by decomposing the measured pressure field into Hankel functions. A comparison of the obtained fields in the frequency domain is given in Figure 3. As may be concluded from this figure, the reconstructed fields match with the original fields. The small variations present are mainly caused by discretization errors.

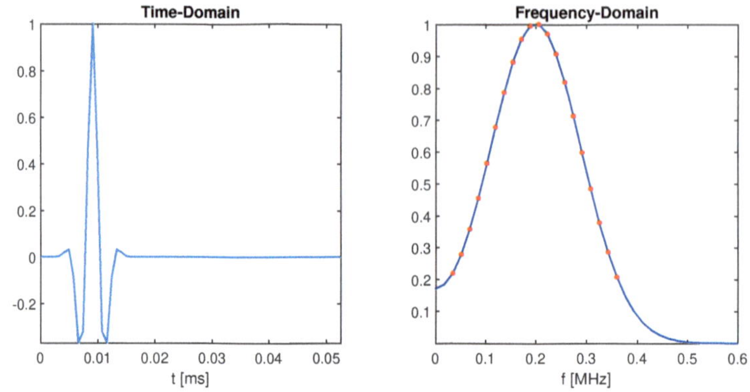

Figure 2: Source excitation profile in time (left) and frequency (right) domain. Red dots show the frequency components that are used for inversion.

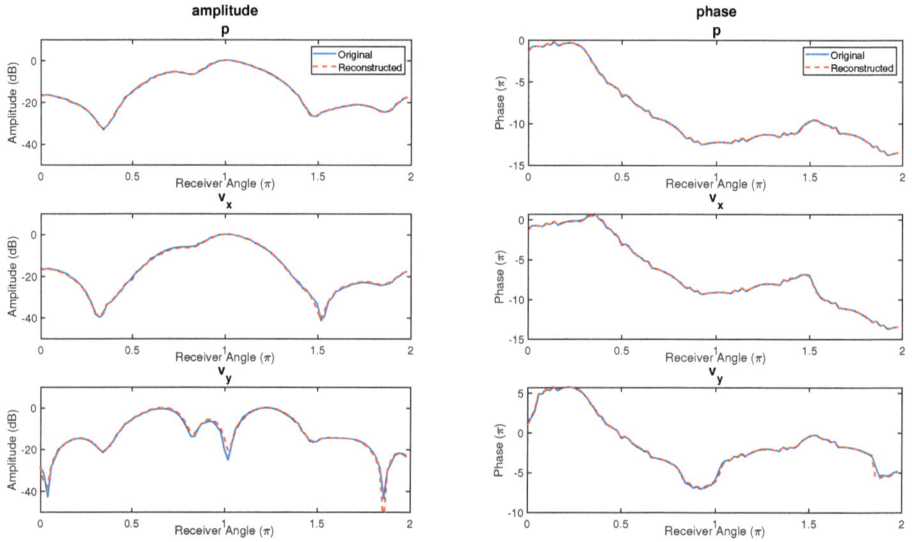

Figure 3: Original (blue) and reconstructed (red dashed) fields in the frequency domain for 0.2 MHz. Amplitude (left column) and phase (right column) of the signals are compared.

3.3 Inversion

A full-vectorial CSI is used as a multi-parameter inversion [8]. This method is applied first on the original pressure and velocity fields to reconstruct density and compressibility contrasts. Next, the method is applied on the reconstructed pressure and velocity fields. The reconstruction results and original contrast profiles are given in Figure 4. Although the reconstruction results are not exactly the same (as expected due to the mismatch between the original and computed fields), the obtained results are promising. It is shown that from

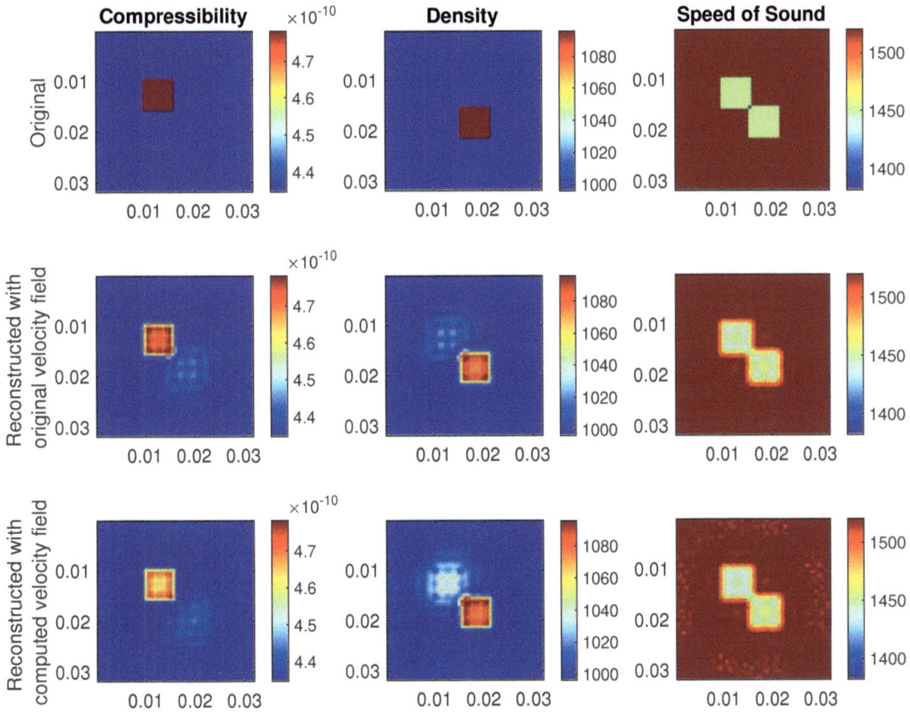

Figure 4: Multi-parameter inversion results. The first row shows the original contrast profiles (compressibility, density and speed of sound). The second row shows reconstruction results when the original velocity field is used in inversion algorithm. The third row shows the reconstruction results when the computed velocity field is used in inversion algorithm.

only pressure field measurements it is possible to compute the velocity field followed by multi-parameter inversion to reconstruct density and compressibility.

4 Conclusion

A full-vectorial CSI is used for multi-parameter inversion. This method reconstructs density and compressibility profiles simultaneously using the measured pressure and velocity fields together. The velocity field is computed from the measured pressure field using Hankel function decomposition. The results obtained with a synthetic example show that the method accurately reconstruct density and compressibility profiles from pressure field measurements only.

References

[1] P.B. Gordon, and S. L. Goldenberg.: Malignant breast masses detected only by ultrasound. a retrospective review. Cancer 76(4) (1995) 626-630

[2] N. Duric, P. Littrup, S. Schimidt, C. Li, O. Roy, L. Bey-Knight, R. Janer, D. Kunz, X. Chen, J. Goll.: Breast imaging with the softvue imaging system: first results. Medical Imaging 8675 (2013)

[3] N. Ruiter, M. Zapf, R. Dapp, T. Hopp, W. Kaiser, and H. Gemmeke.: First results of a clinical study with 3d ultrasound computer tomography. IEEE International Ultrasoinics Symposium (IUS) (2013) 651-654

[4] N. Ozmen, R. Dapp, M. Zapf, H. Gemmeke, N. V. Ruiter, K.W.A. van Dongen.: Comparing different ultrasound imaging methods for breast cancer detection. IEEE Trans. Ultrason., Ferroelectr., Freq. Control 62(4) (2015) 637-646

[5] J. van der Neut, J. T. Fokkema, P. M. van den Berg, M. Zapf, N. V. Ruiter, U. Taskin, and K. W. A. van Dongen.: Ultrasonic Synthetic-Aperture Interface Imaging. IEEE Trans. Ultrason., Ferroelectr., Freq. Control 66(5) (2019) 888-897

[6] J. van der Neut, P. M. van den Berg, J. T. Fokkema and K. W. A. van Dongen.: Acoustic interface contrast imaging. Inverse Problems 34(11) (2018)

[7] M. Pedram, and J. LoVetri.: Evaluation of balanced ultrasound breast imaging under three density profile assumptions. IEEE Transactions on Computational Imaging 3(4) (2017) 864-875

[8] K. W. A. van Dongen, and W. M. D. Wright.: A full vectorial contrast source inversion scheme for three-dimensional acoustic imaging of both compressibility and density profiles. The Journal of the Acoustical Society of America 121(3) (2007) 1538-1549

[9] U. Taskin, J. van der Neut, H. Gemmeke, K. W. A. van Dongen.: Redatuming of 2d wavefields measured on an arbitrary-shaped closed aperture. IEEE Trans. Ultrason., Ferroelectr., Freq. Control (2019)

Quantitative Acoustic Attenuation Imaging of the Breast Using Phase-Insensitive Ultrasound Computed Tomography

D. Sarno[1], C. Baker[1,2], M. Hodnett[1], and B. Zeqiri[1]

[1] *National Physical Laboratory, London, United Kingdom, E-mail: dan.sarno@npl.co.uk*
[2] *King's College London, London, United Kingdom*

Abstract

Ultrasound computed tomography offers a powerful alternative to X-ray mammography; however, reconstruction of the diagnostically important property of acoustic attenuation is exceedingly challenging. Early in the development of ultrasound computed tomography, it was suggested that omnidirectional, phase-insensitive detectors could improve attenuation imaging – the potential of which has been demonstrated through the use of a novel sensor developed at the National Physical Laboratory (Teddington, UK). A prototype phase-insensitive ultrasound computed tomography scanning system has been built to obtain 2D quantitative images of the attenuation of anatomically-sized commercial breast phantoms. The quantitative attenuation images of the breast phantoms obtained using the system compared well to X-ray computed tomography and importantly attenuation measurements matched those made on material samples provided by the phantoms manufacturer.

Future work will systematically evaluate the performance of the imaging system and its ability to differentiate inserts of known sizes and acoustic properties relevant to clinical breast imaging.

Keywords: ultrasound computed tomography, quantitative, attenuation, phase-insensitive

1 Introduction

Ultrasound computed tomography (UCT), while complex, is simply an imaging tool that utilises acoustic property measurements for medical diagnosis. With a number of acoustic properties to measure, it is possible to classify soft tissue types over a multi-dimensional parameter space. From the advent of UCT to the present day, tissue classification through measurements of the two most commonly used parameters – speed of sound (SoS) and attenuation – has been demonstrated [1]. The effectiveness of this approach to tissue classification is enhanced by measuring multiple parameters and through having those measurements be quantitative with low uncertainties.

SoS imaging is most prominent in the field of UCT and – through complex, computationally-intensive reconstruction techniques like full waveform inversion (FWI) – high quality quantitative results have been demonstrated [2, 3, 4]. However, attenuation imaging has had less research focus, in part due to the choice of acquisition hardware used in most existing UCT systems that make quantitative reconstruction of attenuation exceedingly challenging [5].

Our team at NPL approached this problem from a metrology background, applying our expertise in acoustic material characterisation to the field of quantitative attenuation tomography.

1.1 Phase-Insensitive Imaging

When measuring any acoustic property, it is important to use suitable acquisition hardware given the physics of the measurement. For acoustic attenuation, there are a number of considerations that make large, phase-insensitive (PI) sensors a potentially desirable option [6].

Firstly, any medium with variable SoS will cause phase aberration of an acoustic wavefront propagating through it. Aberrated waves incident on a phase-sensitive (PS) detector, with a receiving area of the order of or greater than the wavelength of the acoustic wave, will suffer from phase cancellation across its surface [7, 8, 9]. Increasing the frequency used in measurement compounds the phase cancellation phenomenon. Secondly, SoS variations in a medium cause refraction, deviating acoustic wave propagation off a straight line path. In a through-transmission insertion loss measurement, receivers insufficiently sized to collect all of the acoustic power from the refracted US, overestimate loss [8]. Finally, with soft tissue attenuation following a frequency dependant power law, it is desirable to use higher frequencies to increase attenuation contrast.

The implication of these considerations motivated the "off the mainstream" approach of this research away from PS receivers to PI ones. This was first suggested in the 1970s and 1980s but, at the time, a PI sensor with sufficient sensitivity for clinical use was not available [10, 11].

1.2 The Pyroelectric Sensor

Over the last two decades NPL, in collaboration with Precision Acoustics (Dorset, UK), have developed a novel sensor based on the pyroelectric effect. This is a solid-state power meter with a single, large element receiver area, as shown in Figure 1a. Upon exposure to US, incident acoustic waves pass from water, through a very thin membrane of PVDF and enter a highly absorbing backing later which is well acoustically impedance matched to water. As illustrated in Figure 1b, the ultrasound (US) pressure waves are converted to heat in the absorbing backing layer which is received as a voltage change in the PVDF via the pyroelectric effect. The resultant sensor waveform shape, in the moments just after exposure to US, is related to the rate of change of temperature and the amplitude is proportional to acoustic power [8, 12, 13].

(a) Sensor photographs

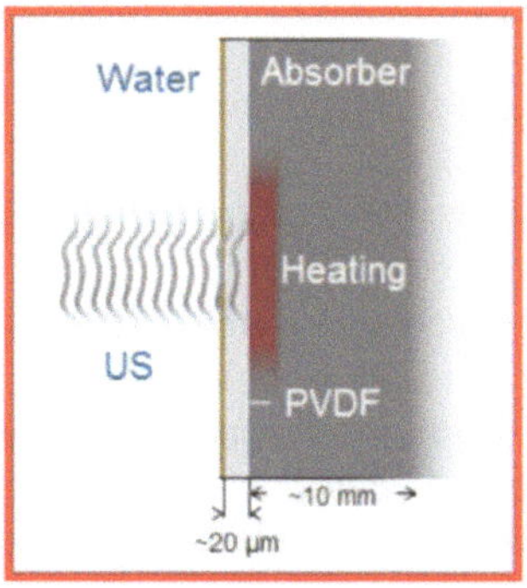

(b) Cross-sectional schematic diagram

Figure 1: (a) Two photographs of the pyroelectric sensor: one face on and the other side on; (b) Schematic diagram of the operation of the pyroelectric sensor, showing US incident from left to right on a cross-section of the sensor layers

Through this mechanism, the novel pyroelectric sensor is ideally suited to attenuation tomography: possessing phase-insensitivity, having a broadband acoustic response, being omnidirectional over a wide range of angles and crucially is sufficiently sized and sensitive for clinical applications.

2 Methodology

A new UCT system was designed and built to investigate attenuation imaging using the phase-insensitive pyroelectric sensor. Efforts primarily focused on quantitative measurement and less on the development of complex reconstruction algorithms. The clinical applicability of phase-insensitive UCT (piUCT) was also considered from the start, with the system designed to accommodate measurements on volunteers in the future.

2.1 Tomographic System

For this early demonstration system, a simple parallel-beam tomographic configuration was used, with each scan comprising of many separate through-transmission insertion loss measurements relative to water [14]. The system utilises 10 mm diameter, 3.2 MHz plane piston transducers (made by Precision Acoustics) which provide long, collimated US beams across the scan field of view (FOV). Fourteen nominally identical transducers make up the transmission array, within which they are arranged horizontally adjacent and emit US in series and parallel to one another. The pyroelectric sensor is positioned opposite the array with a variable separation between the two of 140 mm to 160 mm (see Figure 2d). At their narrowest point, the lateral extent of the beams generated by the transducers are less than the pitch between them, necessitating horizontal movement of the array for insertion loss measurements in the space between adjacent beams. Additionally, the transducers and sensors are mounted on a common arm which rotates the pair around the object from 0° to 180°, in order to complete a tomographic scan. Typical scan sinograms are comprised of 140 lines by 61 angles, totalling 8541 separate insertion loss measurements. Image reconstructions were performed by 4 iterations of the simultaneous algebraic reconstruction technique, SART [15], with a bespoke post-processing technique to correct for acquisition hardware variability [16].

2.2 Phantom Characterisation

The results shown in these proceedings were performed on a commercially-available, multimodality breast biopsy phantom (model 073) and a bespoke urethane phantom, both made by CIRS (Virginia, USA). The model 073 was chosen because of its anatomical accuracy, the presence of randomly positioned inserts and its potential to be used in a comparative exercise with other imaging modalities. The phantom was designed by the manufacturer with material properties that make it suitable to be scanned with magnetic resonance imaging, X-ray computed tomography (XCT) and traditional B-mode US. The 3 mm to 10 mm thick skin material encloses a bulk material which is designed to look heterogeneous under US, containing fibreglass fibres to emulate the appearance of Cooper's ligaments. The bulk encapsulates cyst-like inserts which are anechoic and dense inserts which are hyperechoic under B-mode US. The inserts range in diameter, from 5 mm to 10 mm, with cysts being spherically shaped and the dense masses being both spherical and spiculated in shape. The bespoke phantom differs from the model 073 in having a homogeneous bulk material made of urethane, not having a skin material and containing inserts with different attenuations.

It was important to fully characterise the scanned phantoms for a quantitative evaluation of piUCT images. Working with the phantom manufacturer, we obtained samples of the constituent materials of the phantoms and measured their speed of sound and attenuations under controlled conditions at a range of megahertz frequencies in our materials characterisation facility [17, 18]. Additionally, the phantoms were XCT scanned in order to determine the size, shape and location of the inserts and imaged with traditional hand-held US to confirm the material type of each insert. The attenuation values of the constituent materials, as mea-

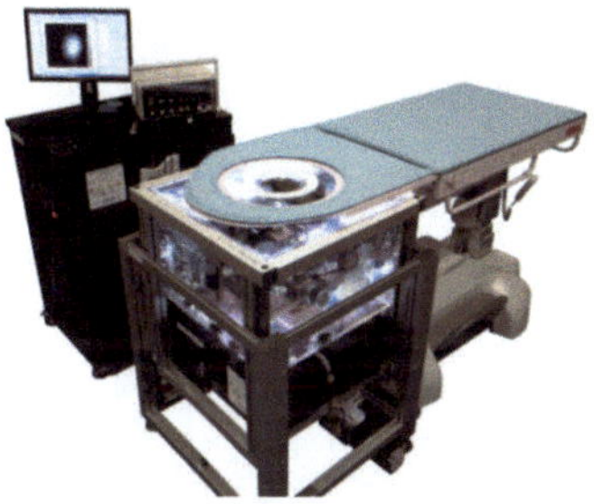

(a) Full piUCT system

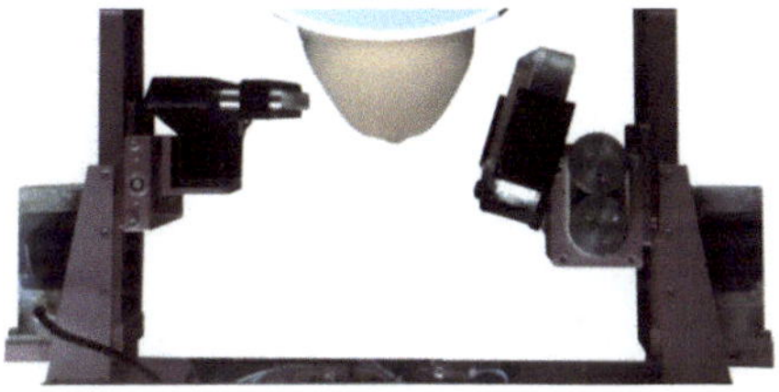

(b) piUCT scanning head and phantom

(c) Pyroelectric sensor and transducers

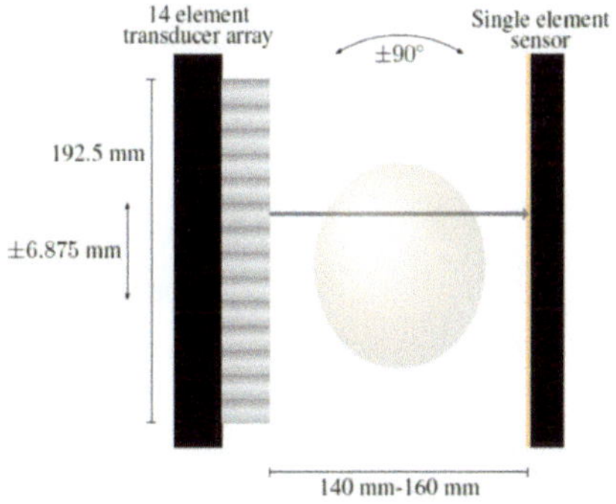

(d) Schematic diagram of tomographic scanning

Figure 2: (a) A photograph of the full piUCT scanning system showing the electronics unit, patient couch and scanning tank under a breast aperture; (b) A side on view of the scanning head comprising of the transducer array on the left opposite the sensor on the right, with the pendant phantom in the centre; (c) The sensor and transducer array, both photographed end on; (d) A schematic diagram of parallel beam tomography from the top down, showing the sequential insertion loss measurements by each transducer and the tomographic motion of the system required to build a scan image

sured by piUCT, were determined by averaging pixel values in regions of the bulk and insert materials.

3 Results

With 3D datasets, many scan parameters and scans of multiple phantoms, understandably there are a lot of results that could be presented. For this simple comparative exercise, we have chosen only to show a couple of representative planes of a commercially-available CIRS phantom and a single plane of a bespoke homogeneous urethane phantom.

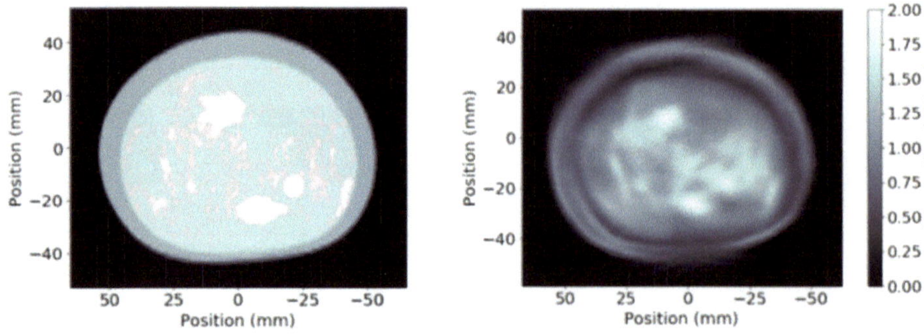

(a) XCT (b) piUCT

Figure 3: Coronal plane 1 of a model 073 breast phantom imaged by XCT and piUCT

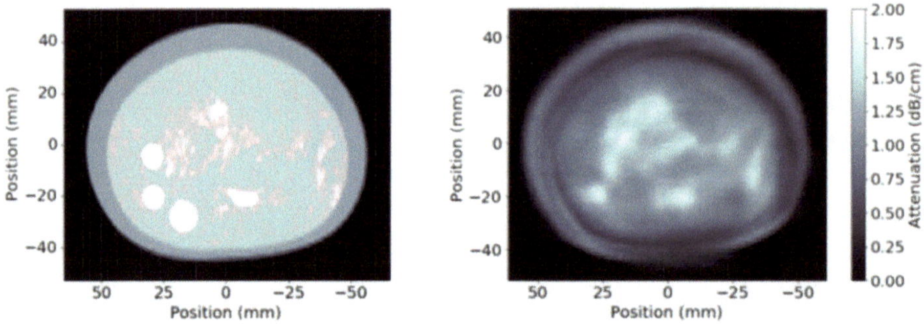

(a) XCT (b) piUCT

Figure 4: Coronal plane 2 of a model 073 breast phantom imaged by XCT and piUCT

The first plane of the model 073 phantom, Figure 3, contains two spiculated dense masses (one at the 11 o'clock position and one at the 5 o'clock position) and one spherical dense mass (at the 4 o'clock position), as clearly seen in Figure 3a.

The second plane of the model 073 phantom, Figure 4, is notable for containing two spherical cystic masses. These are the highest and lowest positioned inserts of the trio located at the 7 o'clock position of Figure 4a. The middle of the three inserts at the 7 o'clock position and the insert located at the 6 o'clock position are a spherical dense mass and a spiculated dense mass, respectively. The cross-sectional size of the second plane, Figure 4, is greater than that of the first, Figure 3, due to the tapered shape of the phantom.

The scan plane of the urethane phantom, Figure 5, contains two dense masses with the left insert being spherical and the right being spiculated. This bespoke urethane phantom does not have a skin layer like that seen in the model 073 phantom. Also presented is a profile of the pixel values of the XCT and piUCT images at the position denoted by the horizontal red line in Figure 5a and Figure 5b.

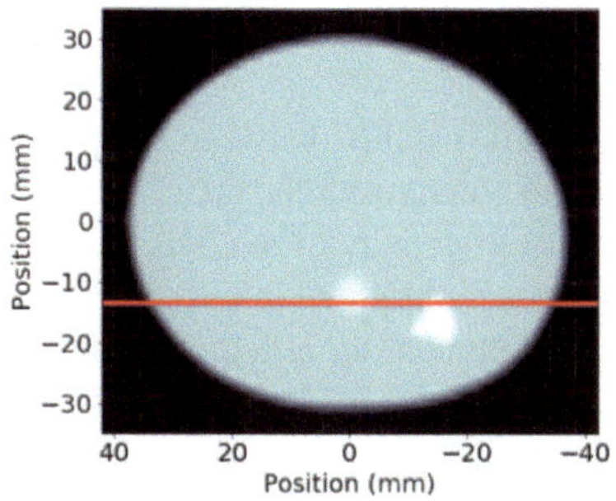

(a) XCT

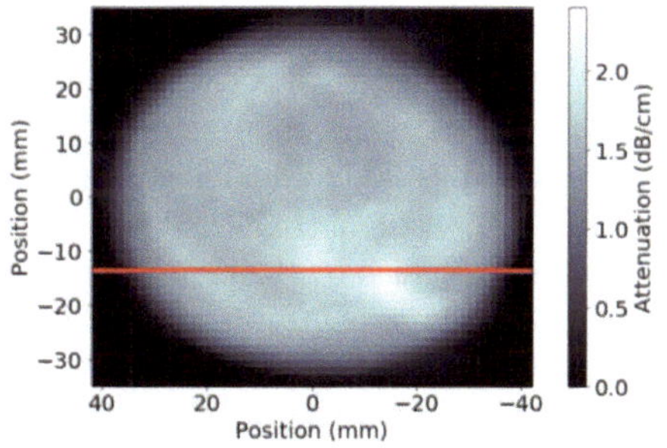

(b) piUCT

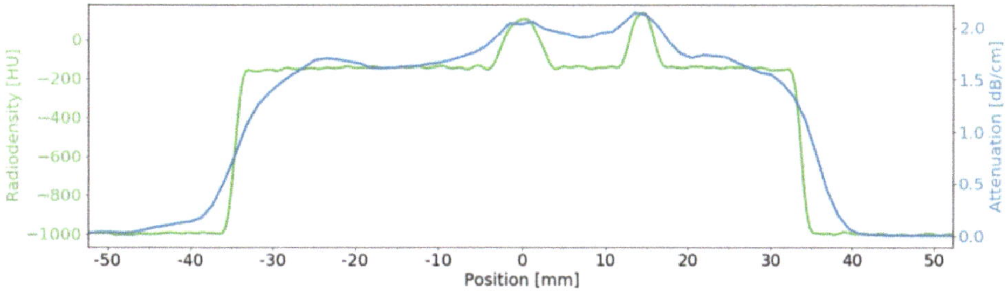

(c) A radiodensity profile against acoustic attenuation profile of the horizontal lines in (a) and (b)

Figure 5: Coronal plane of a homogeneous urethane breast phantom imaged by XCT and piUCT with a horizontal profile, crossing two highly attenuating inserts, as measured by each imaging modality

Finally, Table 1 displays the attenuation values measured by the materials characterisation facility on material samples in comparison to the attenuation values extracted from the piUCT scan images. The four constituent materials of the model 073 are presented in addition to the two constituent materials of the bespoke urethane phantom.

Phantom	Material	True Attenuation [dB cm^{-1} at 3.2 MHz]	piUCT Attenuation [dB cm^{-1} at 3.2 MHz]
Model 073	Cyst	0.9 ± 0.1	1.0 ± 0.1
	Dense	2.2 ± 0.1	1.9 ± 0.2
	Bulk	1.0 ± 0.1	1.0 ± 0.3
	Skin	0.7 ± 0.1	0.7 ± 0.3
Bespoke	Urethane	1.6 ± 0.1	1.6 ± 0.1
	Dense	2.5 ± 0.1	2.3 ± 0.2

Table 1: Quantitative acoustic attenuation values of the constituent materials of the phantoms as measured on material samples in the lab, detailed in [18], and as measured *in situ* by the piUCT scanning system

4 Discussion

4.1 Spatial Comparison

In general, the skin and bulk interior shape and size are well matched between the piUCT scan image and the XCT scan image. One of the more notable features of the piUCT scans of the model 073 phantom (Figure 3b and Figure 4b) is the heterogeneity of the bulk material. This can be explained by the presence of the fibreglass fibres highlighted in red in the XCT scan images of both planes. These are likely to be highly attenuating, at least in part, due to the presence of a coating of microbubbles introduced in the manufacturing process – a theory postulated by the manufacturer. Fibre density and position match well between piUCT scan images and highlighted XCT images, with fibres that are isolated and follow along the coronal plane of the piUCT scan being resolved more clearly.

The dense inserts in both planes of the model 073 phantom are well resolved and positioned in the piUCT scan images. However, the spiculated and spherical shapes of the dense inserts are difficult to distinguish in the piUCT scan images, with the XCT scans possessing superior spatial resolution. The cyst-like insert are unresolved against the background material in Figure 4b. This low attenuation contrast of the cystic masses against the background material is confirmed by the similarity of the attenuation values, approximately $1\,\mathrm{dB\,cm^{-1}}$, of the two materials as measured on samples in the lab, see Table 1.

In the bespoke urethane phantom scans, there is good agreement in the size, shape and locations of the inserts and the homogeneous bulk material between the piUCT scan image and the XCT scan image. The lower spatial resolution of piUCT relative to XCT is evident in horizontal profiles of each, as seen in Figure 5c, with the beamwidths of the former being larger than the latter. The difference in shape between the two inserts is more clear in this bespoke phantom likely due to the homogeneous bulk material offering greater contrast and distinction in the area around the inserts.

4.2 Attenuation Comparison

Evaluating the attenuation values in Table 1, it is clear that the piUCT, *in situ* derived, measurements of acoustic attenuation match well to those measured under controlled conditions on blocks of material samples in the lab. This is an encouraging result given that the reconstruction algorithm used – 4 iterations of SART – only took approximately 1 s to complete on a lab PC and whose model is simple and assumes infinitely narrow, straight rays.

The larger discrepancies between the piUCT derived insert attenuations and the ground truth values are likely due to the spatial averaging phenomenon from a finite beam distribution of the US source, clearly demonstrated in Figure 5c. This can be improved by using more collimated beams with acoustic power confined to a narrower volume over a longer distance of travel through the scan FOV.

The greater uncertainty of attenuation of the larger materials of the model 073 phantom – skin and bulk in Table 1 – is also notable. This is likely a result of a refraction artefact between the skin and background materials, due to a discontinuous change in SoS at the boundary of materials, and the presence of the fibreglass fibres in the background material. The former, reconstructing as a reduction in attenuation at the material interface, likely as a result of the US following a curved path through the lower SoS and attenuating skin rather than passing through the higher attenuating bulk, as the straight ray model of SART assumes. The latter, reconstructing as a heterogeneous background with a greater variation in attenuation due to varying amount of scattering microbubbles coating the fibres and out of plane effects from a complex 3D fibre structure.

Reduced attenuation variability can be seen in the urethane phantom bulk material, owing to its homogeneous construction without the presence of scattering fibres. The small increase in the measured attenuation at the −25 mm and 25 mm positions of Figure 5c is likely due to the presence of a ring artefact, coincident with the edge of the phantom, as a result of acquisition hardware variability and is the subject of current research [16].

4.3 Future Work

A future aim of this piUCT research is to make our tomographic measurements definitively quantitative, with measurement values paired to uncertainty values. The implication, in the field of quantitative imaging in which each pixel is effectively a measurement, is for scan images to be paired with associated uncertainty images. For image reconstruction, the algorithms that transform the individual acquisition measurements – often represented together in a sinogram – to a scan image are often too complex to propagate errors through purely algebraically. An alternative approach to error propagation is through Monte Carlo simulations, where measurement variability is simulated and images reconstructed many times in order to determine the probability distribution and standard deviation of each pixel. This can mean performing many hundreds of reconstructions which, with our use of simple reconstruction algorithms, is not computationally or time intensive, only taking minutes to perform on regular PC hardware.

Another area of improvement is to increase the complexity of the reconstruction measurement model to account for some of the physical mechanisms omitted in SART. The main two physical mechanisms being: finite beam distributions and refraction effects. The former is in initial development with a Beam Area Algebraic Reconstruction Technique (BAART) showing some early signs of improvement over SART on a quantitative measurement basis. The technique utilises a priori knowledge of the nominal beam distribution of transducers on the piUCT scanner in the reconstruction model. The latter is more of a long-term goal, with a bent ray reconstruction likely requiring knowledge of the speed of sound distribution of the medium which is currently not measured.

With regards to SoS imaging, although the piUCT system is designed for power measurements and attenuation imaging, the pyroelectric effect – upon which the sensor is based – is irreducibly coupled to the piezoelectric effect. In practice this means that our sensor not only responds to acoustic power, but also acoustic pressure, with the higher frequency component of the signal currently filtered from the lower frequency pyroelectric signal. In the future, we hope to make use of this piezoelectric sensor response to derive SoS information to aid our quantitative attenuation reconstructions.

5 Conclusion

In these proceedings, we have presented an "out of the mainstream" approach to quantitative attenuation imaging through the use of a novel phase-insensitive pyroelectric sensor developed at the NPL. Our aim was to focus on measurement quality rather than development of complex reconstruction algorithms. We were able to generate quantitative images that matched well to the ground truth in terms of both the spatial distribution of pixels and, most importantly, on quantitative attenuation measurements of the constituent phantom materials. With the collective aim of UCT to identify tissue through measurements of acoustic properties, it is imperative to quantitatively measure multiple parameters with low uncertainty. Given that SoS and attenuation parameters have different physical considerations of their measurement, it is not unreasonable to think that each would require dedicated measurement hardware and that a combination of approaches would be required in a complete, quantitative UCT system.

References

[1] J. Greenleaf, R. Bahn: Clinical Imaging with Transmissive Ultrasonic Computerized Tomography. IEEE Transactions on Biomedical Engineering BME-28(2) (1981) 177-185

[2] N. Duric, P. Littrup, C. Li, O. Rama, L. Bey-Knight, S. Schmidt, J. Lupinacci: Detection and Characterization of Breast Masses with Ultrasound Tomography: Clinical Results. Proceedings of SPIE - The International Society for Optical Engineering 7265 (2009)

[3] M. André, C. Barker, N. Sekhon, J. Wiskin, D. Borup, K. Callahan: PreClinical Experience with Full-Wave Inverse-Scattering for Breast Imaging. Acoustical Imaging, Dordrecht, Springer (2009) 73-80

[4] N. Ruiter, M. Zapf, R. Dapp, T. Hopp, H. Gemmeke: First In Vivo Results with 3D Ultrasound Computer Tomography. IEEE International Ultrasonics Symposium (2012)

[5] M. Pérez-Liva, J. Herraiz, J. Udías, E. Miller, B. Cox, B. Treeby: Time Domain Reconstruction of Sound Speed and Attenuation in Ultrasound Computed Tomography Using Full Wave Inversion. Journal of the Acoustical Society of America 141(3) (2017) 1595–1604

[6] J. Klepper, G. Brandenburger, J. Mimbs, B. Sobel, J. Miller: Application of Phase-insensitive Detection and Frequency-Dependent Measurements to Computed Ultrasonic Attenuation Tomography. IEEE Transactions on Biomedical Engineering BME-28(2) (1981) 186-201

[7] J. Klepper, G. Brandenburger, L. Busse, J. Mille:, Phase Cancellation, Reflection, and Refraction Effects in Quantitative Ultrasonic Attenuation Tomography. IEEE Ultrasonics Symposium (1977)

[8] B. Zeqiri, C. Baker, G. Alosa, P. Wells, H. Liang: Quantitative Ultrasound Computed Tomography Using Phase-Insensitive Pyroelectric Detectors. Physics in Medicine and Biology 58(15) (2013) 5237-5268

[9] J. Bamber: Attenuation and absorption. Physical Principles of Medical Ultrasonics (1986) 93–166

[10] J. Heyman, J. Cantrell: Application of an Ultrasonic Phase Insensitive Receiver to Material Measurements. IEEE Ultrasonics Symposium (1977).

[11] D. Fitting, P. Carson, J. Giesey, P. Grounds: A Two-Dimensional Array Receiver for Reducing Refraction Artifacts in Ultrasonic Computed Tomography of Attenuation. IEEE Transactions on Ultrasonics, Ferroelectrics and Frequency Control, 34(3) (1987) 346–356

[12] B. Zeqiri, P. Gélat, J. Barrie, C. Bickley: A Novel Pyroelectric Method of Determining Ultrasonic Transducer Output Power: Device Concept, Modeling, and Preliminary Studies. IEEE Transactions on Ultrasonics Ferroelectrics and Frequency Control 54(11) (2007) 2318-2330

[13] B. Zeqiri, J. Barrie: Evaluation of a Novel Solid-State Method for Determining the Acoustic Power Generated by Physiotherapy Ultrasound Transducers. Ultrasound in Medicine and Biology 34 (9) (2008) 1513-1527

[14] C. Baker , D. Sarno, R. Eckersley, B. Zeqiri: Pulse Pileup Correction of Signals From a Pyroelectric Sensor for Phase-Insensitive Ultrasound Computed Tomography. IEEE Transactions on Instrumentation and Measurement 68(10) (2019) 3920-3931

[15] A. Andersen, A. Kak: Simultaneous Algebraic Reconstruction Technique (SART): A Superior Implementation of the ART Algorithm. Ultrasonic Imaging 6(1) (1984) 81-94

[16] C. Baker , D. Sarno, R. Eckersley, B. Zeqiri: Ring Artefact Correction for Phase-Insensitive Ultrasound Computed Tomography. IEEE Transactions on Ultrasonics Ferroelectrics and Frequency Control *Early Access* (2019)

[17] S. Rajagopal, N. Sadhoo, B. Zeqiri: Reference Characterisation of Sound Speed and Attenuation of the IEC Agar-Based Tissue-Mimicking Material Up to a Frequency of 60 MHz. Ultrasound in Medicine and Biology 41 (1) (2015) 317–333

[18] R.Baêsso, R.Costa-Felixa, P. Miloro, B. Zeqiri: Ultrasonic Parameter Measurement as a Means of Assessing the Quality of Biodiesel Production. Fuel 241 (2019) 155-163

Ultrasound Transducer Identification enables High-resolution Full-waveform Inversion

C. Cueto[1], L. Guasch[1], J. Cudeiro[1], O. Calderon Agudo[1], M. Warner[1], and M.-X. Tang[1]

[1] *Imperial College London, London, United Kingdom, E-mail: c.cueto@imperial.ac.uk*

Abstract

Background

Full-waveform inversion (FWI), a technique developed in geophysics, has recently been applied to produce high-resolution and quantitative reconstructions in the context of medical ultrasound imaging. However, even though FWI has shown great promise, realistic acquisition conditions introduce a series of uncertainties that cannot be fully considered using existing schemes. This could lead to a degradation of resolution and accuracy, and even make the solution of the inverse problem impossible. One such sources of uncertainty is related to the behaviour of transducers. Existing schemes make strong assumptions on their behaviour, frequently considering them to be point-like, while also neglecting their inherent electro-acoustic response. Here we extend traditional FWI algorithms by using a technique for the characterisation of transducers that ensures a highly accurate modelling of their response. Subsequently, we explore the impact that uncertainties in transducer behaviour have on the resulting reconstructions.

Methods

The proposed technique extends traditional FWI schemes through the addition of a secondary optimisation problem whose aim is to find a suitable transducer model from measurements while honouring the full wave equation. This leads to transducer models that are well suited for FWI, as well as other techniques such as optoacoustic tomography, in which close agreement between experimentally acquired and numerically modelled data is needed. The technique is validated on experimental measurements acquired using disk transducers of different diameters, with and without lenses. We then compare the results in terms of errors in magnitude and phase with those obtained using common transducer modelling approaches: neglected response, deconvolution and holography/backpropagation. The impact on FWI of uncertainties in transducer behaviour is explored using a numerical phantom of the human head by comparing the quality of the reconstructions when transducer response is correctly or incorrectly modelled.

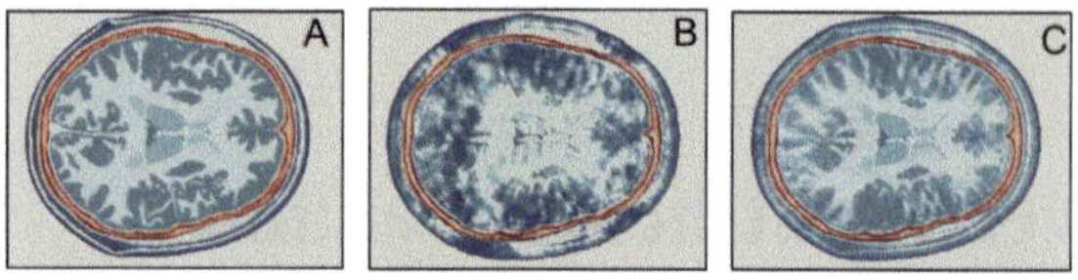

Figure 1. Numerical head phantom (**A**) whose FWI reconstruction is degraded by uncertainties in transducer behaviour (**B**); reconstruction quality is recovered when introducing correct transducer models (**C**). Comparison of different methods for transducer identification applied to a 16 mm disk transducer in terms of magnitude error (**D**) and phase error as percentage of a cycle (**E**), calculated over 16762 spatially independent samples of the wavefield.

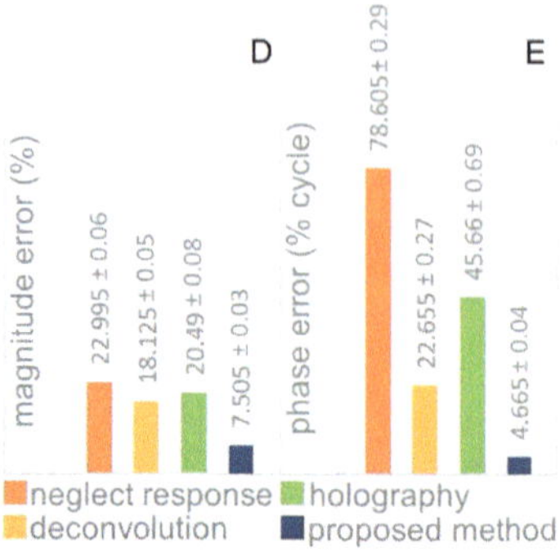

Results

Figure 1-A to C suggest that a high degree of certainty on the transducer response is needed for FWI under realistic acquisition conditions. When compared to commonly employed transducer models, Figure 1-D and E confirm that the proposed technique can achieve superior transducer identification performance.

Overcoming cycle-skipping in full-waveform inversion of ultrasound data

O. Calderon Agudo[1], G. Stronge[1], T. Robins[2], and L. Guasch[1]

[1] *Imperial College London, Department of Earth Science and Engineering, London, United Kingdom, E-mail: ocl4@ic.ac.uk*
[2] *Imperial College London, Department of Bioengineering, London, United Kingdom*

Abstract

Full-waveform inversion (FWI) tomographic algorithms hold great potential for medical imaging; they are able to image sub-millimetre morphological features and provide quantitative information of tissue properties such as speed-of-sound. Standard FWI algorithms seek a model of physical properties that minimise the L2 norm of the difference between the observed data acquired experimentally and numerically modelled data. Due to the non-linearity and non-convexity of the inverse problem, FWI requires low-frequency data or accurate starting models in order to converge to the correct solution. Misconvergence is usually caused by the so-called cycle-skipping effect, in which the observed and modelled data are out of phase by more than half a cycle, yielding incorrect recovered models. Issues arise because most commercially available ultrasound transducers lack low-frequencies. Additionally, starting models are not available for highly-heterogenous or complex datasets, such as those containing hard tissues. Thus, FWI does not always produce reliable results. Here we explore the application of two methodologies developed in the context of geophysics that are resilient to cycle-skipping: adaptive waveform inversion (AWI) and envelope inversion (EI). They are tested on three different datasets: 1) the 2019 USCT 2D blind dataset, 2) a laboratory dataset acquired for a small soft-tissue brain phantom and 3) a dataset generated synthetically with low-frequency transducers for an in-silico model representative of the brain that is surrounded with a hard skull layer. We demonstrate that the use of such techniques enables the recovery of accurate models, even when using high-frequency transducers or in the presence of hard tissues.

Keywords: Full-waveform inversion, adaptive waveform inversion, envelope inversion

1 Introduction

Accurate and high-resolution methods for imaging both soft and hard tissues are critical for the detection and diagnosis of a wide range of pathologies. The predominant techniques used for imaging targets containing hard tissues are magnetic resonance imaging (MRI) and x-ray computed tomography (CT) [1, 2]. Despite the widespread successful application of these methods, they suffer from a variety of disadvantages. MRI is contraindicated if magnetic foreign bodies are present and irremovable; x-ray CT exposes the patient to harmful ionising radiation and is typically lower resolution, with poor soft-tissue contrast; finally, both MRI and x-ray CT machines are immobile and expensive. Contrarily, ultrasound methods are safe, portable, fast and unexpensive. However, they typically have a significantly lower spatial resolution and are unable to image through hard tissues, thus prohibiting their application for imaging beneath bone.

Ultrasound computer tomography (USCT) has established itself as an alternative method for imaging soft tissues with ultrasound, particularly for breast cancer diagnosis [3]. This is advantageous over the conventional use of mammography as it does not expose patients to harmful ionising radiation. However, most implementations rely on ray theory and, hence, the spatial resolution of such methods is usually restricted to the first Fresnel zone [4]. On the other hand, full-waveform inversion (FWI) methods can overcome this limitation and produce 3D images of the breast at the sub-wavelength scale at the expense of a higher computational cost [5]; this is a vast improvement on the resolution ability of ray-based techniques. FWI is able to achieve this by using the wave equation to predict the entire waveforms in the data, thereby incorporating the full physics of finite-frequency acoustic waves [6]. This also means that FWI can potentially be used for imaging through hard tissues as wave distortions at interfaces are accurately modelled, thus enabling imaging of inaccessible regions such as the brain.

The primary limitation of FWI is that it converges to a local minimum if the observed and predicted data differ by more than half a cycle, as illustrated in Figure 1. This results in a recovered model that is usually no better, if not worse, than the starting model [6]. This is a well known phenomenon, commonly referred to as cycle skipping [7], and it is a consequence of the non-linear relationship between model parameters and pressure in the wave equation [6].

There are two general strategies to mitigate cycle-skipping. The first is to ensure that the closest minimum in the FWI objective function is the global minimum. This can be accomplished either by constructing a good starting model, or by acquiring low-frequency data and employing a multi-scale approach, where the inversion commences at low frequencies with higher frequencies being gradually introduced [6]. The second strategy is to change the optimisation problem. In geophysics, various ideas have been proposed to do this: adaptive waveform inversion (AWI) uses Wiener filters to match the observed data to the predicted data, and updates the model by optimising the Wiener filter coefficients at each iteration [7]; whilst envelope inversion (EI) minimises the difference between the envelopes of the pre-

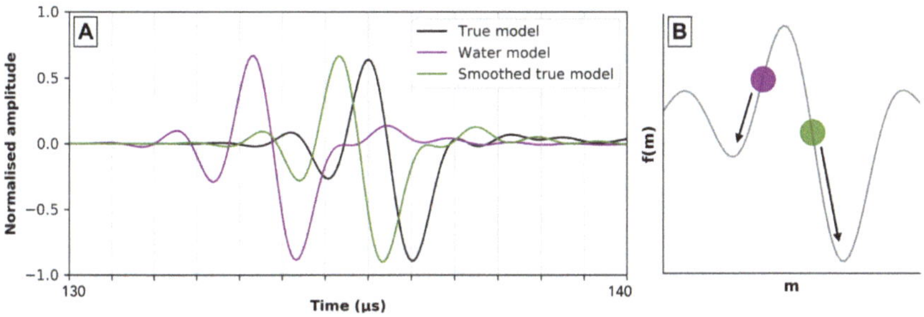

Figure 1: Synthetic data generated using a true (USCT data challenge) model, water model and smoothed true model (left). The water model data is cycle-skipped with respect to the true model, whereas the smoothed true model data is not. Schematic diagram of the FWI objective function illustrating convergence of water model to a local minimum, whereas the smoothed starting model will converge towards the global minimum (right).

dicted and observed waveforms, rather than the waveforms themselves [8]. Other methods include using an objective function based on optimal transport theory [9], or generating intermediate datasets [10]. Most of these incur significant additional computational costs and/or complexities, with the exception of AWI and EI. Hence, we only consider AWI and EI in this paper.

In the following section, we first review the basic theory of FWI, AWI and EI, emphasising the differences in their objective functions and how these relate to cycle-skipping. We then apply the proposed algorithms to three datasets, each of which presents a different form of cycle-skipping.

2 Methodology

FWI is a method that was originally developed to image the subsurface in geophysics by solving a local optimisation problem [6]. It attempts to recover a high resolution model, typically of acoustic velocity, by learning to accurately predict the entire waveforms present in some observed data. Here we outline the fundamental theory required to understand FWI within the context of medical imaging, before introducing the different objective functions that are minimised in AWI and EI.

An acoustic velocity model $\mathbf{m}$ can be used to generate predicted data $\mathbf{p}$ by solving the forward problem $G(\mathbf{m}) = \mathbf{p}$, where G is a non-linear operator that represents a numerical implementation of the (acoustic) wave equation. The aim of FWI is to solve the inverse problem $\mathbf{m}' = G^{-1}(\mathbf{d})$, where $\mathbf{m}'$ is an unknown model and $\mathbf{d}$ is some observed data. However, solving the inverse problem directly is not possible as G has no formal inverse [6]. The FWI algorithm proceeds by using a local iterative inversion scheme, invoking the Born approximation to make a series of linearised updates to the model that minimise the following least-squares norm f_{FWI}:

$$f_{FWI}(\mathbf{m}) = \frac{1}{2}||\mathbf{p} - \mathbf{d}||_2^2 \tag{1}$$

In order to calculate the model update, the gradient of the objective function with respect to the model is first calculated using the adjoint-state method [6]. For a single source, this involves back-propagating the residual between the observed and forward modelled predicted data, and for each time-step, cross-correlating the back-propagated residual wavefield with the forward wavefield across the entire model. The model update $\delta\mathbf{m}$ is then given by:

$$\delta\mathbf{m} = -\mathbf{H}^{-1}\frac{\partial f_{FWI}(\mathbf{m})}{\partial\mathbf{m}} \tag{2}$$

where $\mathbf{H}$ is the Hessian operator, containing second-order differentials of the objective function with respect to the model. In practice, calculating the inverse of the full Hessian is unfeasible, thus only the diagonal values of an approximate Hessian are used [6].

In AWI, rather than minimising the difference between the predicted and observed data, the ratio of the predicted and observed data is forced to be equal to one [7]. This is done by defining a Wiener filter $\mathbf{w}$ that, when convolved with the observed data $\mathbf{d}$, provides an optimal least squares match to the predicted data $\mathbf{p}$. The AWI objective function is then defined as:

$$f_{AWI}(\mathbf{m}) = \frac{1}{2}\frac{||\mathbf{Tw}||_2^2}{||\mathbf{w}||_2^2} \tag{3}$$

where $\mathbf{T}$ is referred to as the annihilator, which increases monotonically away from zero-lag, thereby penalising energy at larger lags and forcing the Wiener filter $\mathbf{w}$ to be a zero-lag delta function [7].

EI is conceptually more similar to FWI than AWI. Instead of minimising the difference between the predicted and observed data, EI minimises the difference between their envelopes [8]. This helps to mitigate cycle skipping as the envelopes contain low frequency information. The EI objective function is defined as:

$$f_{EI}(\mathbf{m}) = \frac{1}{2}||E(\mathbf{p})^{\alpha} - E(\mathbf{d})^{\alpha}||_2^2 \tag{4}$$

where E represents the envelope operator and α is a power coefficient that can be adjusted to precondition the data [8].

In the next section, we explore the advantages and limitations of using AWI and EI over FWI for inverting three different ultrasound datasets.

3 Results

We now evaluate the performance of FWI, AWI and EI on three datasets: 1) the 2019 USCT blind data challenge, 2) a laboratory dataset acquired with standard ultrasound probes for a small brain phantom and 3) a synthetic dataset for a human-sized brain in-silico phantom surrounded with a hard-tissue layer. These datasets present different challenges. The first is a synthetic dataset that has been generated using an unusual wide-bandwidth source and a soft-tissue breast model, for which the true model used to generate the data is unknown. The first step was to recover an accurate model with standard FWI. We then further complicated the dataset by filtering its spectrum with that of a standard ultrasound probe. The second, a soft brain phantom, was acquired using standard probes in the laboratory. This dataset is challenging as it has realistic noise levels, a narrower data bandwidth and the source positions were not calibrated. Finally, the last dataset was again synthetically generated, but now the true model of speed of sound is known and contains a hard-tissue layer. The presence of the latter leads to cycle-skipping of transmitted arrivals if no prior information of the skull is used, resulting in inaccurate recovered models with FWI. Thus, the use of functionals resilient to cycle-skipping is necessary on these datasets.

3.1 USCT data challenge

First, we utilized the 2D ring aperture blind test from the 2019 USCT data challenge to evaluate the performance of the proposed inversion algorithms under different circumstances. As the true model is unknown, we initially inverted the data to recover a model of speed-of-sound, which was then used to perform various tests. Our inversion strategy consisted of minimal pre-processing of the data by just resampling in time, deleting bad traces or shot gathers, and extracting a source wavelet from the water dataset as in [5].

The data contained usable energy at relatively low frequencies, i.e. below $200kHz$. This meant that avoiding cycle-skipping effects in FWI, by using a multi-scale approach that inverted progressively higher frequency bands of the data using a homogenous water starting model, was straightforward. We inverted the blind dataset from $200kHz$ up to $3MHz$ in a total of 104 iterations, using the inversion algorithm described in [6], but without calibrating source positions as these were considered to be fixed. This resulted in the recovered speed-of-sound model of a breast section displayed in Figure 2a. Five different regions can be distinguished in this model: a background medium with the speed-of-sound of water, an outer fast skin layer, two inner structures representing glandular and fat tissues, and a faster localised region that may represent a tumour. This model is now referred to as the true model, and is used to generate data again with the extracted source wavelet, and the same number of transducers as in the data challenge. Hence, an optimal inversion result from this data should lead to a model similar to that in Figure 2a. We computed the relative speed-of-sound error of each model with respect to the former, and we displayed this value on the corresponding image.

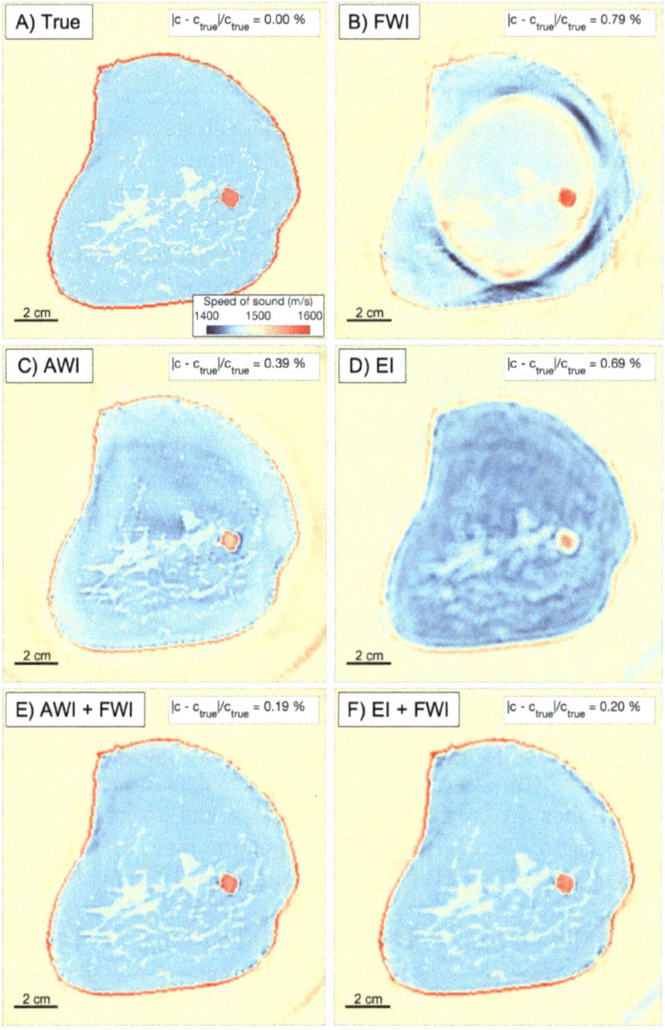

Figure 2: The recovered speed-of-sound model from the USCT data challenge in A) is used as a reference to generate synthetic data, which are inverted using a homogeneous water starting model and FWI (52 iterations), AWI and EI (20 iterations each) with a starting frequency of 500 kHz, yielding the models in panels B) to D), respectively. The models in C) and D) are then used as starting models for FWI, yielding the models in panels E) and F). Cycle-skipping artefacts present in B) are mitigated in panels C) to F).

The first test performed with this data shows that cycle-skipping artefacts affect the resulting recovered models with FWI if the multi-scale approach starts at a relatively high frequency of $500kHz$ and a homogeneous water starting model is used. In this case, the predicted and observed data differ by more than half a cycle, leading to cycle-skipping artefacts in the recovered model. These inaccuracies are visible in Figure 2b: the outline of the breast and tumor are clearly recovered, but the outer skin layer is not, and more importantly, the inner structure of the breast is lost in several regions and is substituted by high-speed artefacts due to cycle-skipping effects. Instead, Figures 2c and 2d show that just 20 iterations of AWI or EI starting at the same frequency are enough to mitigate cycle-skipping effects, resulting in models that contain all regions of the reference breast model but at a low resolution. Consequently, there is a decrease in the relative error when compared to that in Figure 2b for both models, which is moderately more significant for Figure 2c than for Figure 2d.

The models in Figures 2c and 2d are then used as starting models for a subsequent inversion using FWI up to $2MHz$ and starting at the same frequency, resulting in the models in Figures 2e and 2f, respectively. When compared to the model in Figure 2b, these models are more accurate and do not contain visible artefacts because AWI and EI are more resilient to cycle-skipping. We note that the differences between the models in Figures 2e and 2f are minimal, and the resolution of the different features would further increase when extending the inversion up to $3MHz$; this would, however, increase the overall computational cost.

We then modified the data generated from the model in Figure 2a to make it more realistic by filtering their bandwidth and adding white noise. The data was filtered with the normalised spectrum measured in the laboratory for a P4-1 linear ultrasound probe transmitting a 3-cycle toneburst at $1.4MHz$. We added white noise to the data so that the signal-to-noise ratio (SNR) is $35dB$ except for a few receivers (with numbers 10, 120 and 200), in which we increase the noise to an SNR of $2dB$ to simulate defective transducers. Figures 3a and 3b show a shot gather of the original and modified datasets, whereas Figure 3c displays the amplitude spectra of the original and modified data contained within the black box in Figures 3a and 3b. The bandwidth of the modified data is reduced, and the level of noise has increased.

Due to the much narrower bandwidth of the data, inversions need to start at a much higher frequency. In Figure 4, we used AWI and EI both followed by FWI, and also FWI by itself on the modified dataset, starting from a homogeneous water model and a frequency of $830kHz$. The recovered models are displayed in Figures 4b to 4d.

FWI alone results in a model in which the tumour and the internal features of the breast are not accurately recovered. AWI followed by FWI, however, is able to recover the tumour location, characteristics and some of the internal structure, despite the added noise and the much narrower data bandwidth. Despite this, the lack of lower frequencies prevents AWI from obtaining a more accurate model that contains the skin, and some artefacts inside the breast due to cycle-skipping are also visible. Although AWI is resilient to cycle-skipping, its performance for breast imaging deteriorates when lower frequency data are not available. We therefore recommend the use of lower frequency transducers for optimal results, com-

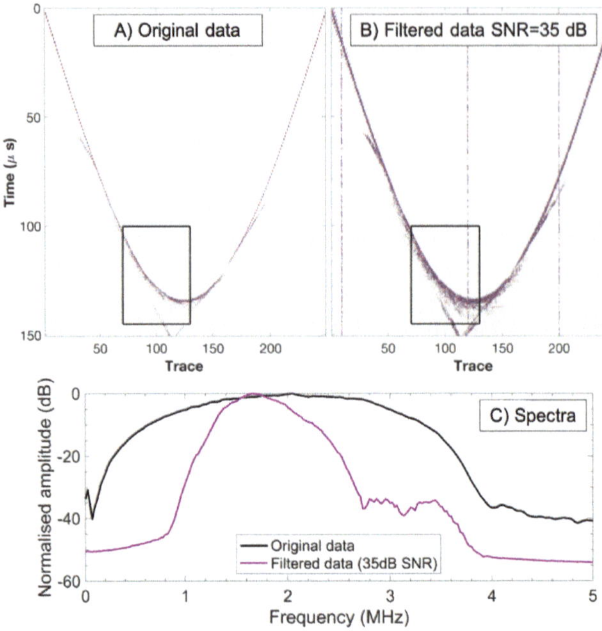

Figure 3: The original gathers from the USCT data challenge (panel A) are made more realistic by filtering the data with the measured signal of a standard P4-1 probe measured in the laboratory, yielding the data in panel B, in which white noise has also been added ($SNR = 35dB$ except for a few traces with $SNR = 2dB$). Panel C displays the spectrum of the original and modified data within the box in panels A and B.

bined with the use of AWI. This is especially important for accurate and robust breast cancer diagnosis. In contrast, EI followed by FWI does not result in an accurate and representative model because the average speed of sound is not correct, which leads to the model in Figure 4d. Thus, FWI and AWI are used from here onwards, and we suggest further research into alternative formulations of EI that improve its resilience to cycle-skipping for more realistic datasets.

3.2 Laboratory brain phantom data

Both FWI and AWI are now tested on a dataset acquired in the laboratory of a small phantom that is $7.5\,cm$ on its longest side, and $5.5\,cm$ on its shortest side. The phantom is made of biodegradable polyvinyl alcohol (PVA) and represents a structure similar to that of a brain. A picture of a cross-section of the phantom is shown in Figure 5a. The phantom is a vertical extension of this cross-section in 3D to avoid large vertical variations, so that it is effectively 2.5D. It consists of three different regions, all made with PVA, that have different speeds of sound due to their composition. These simulate different regions in the brain. The middle region is hollow, and is filled with water when inserted in a water tank. Ultrasound data

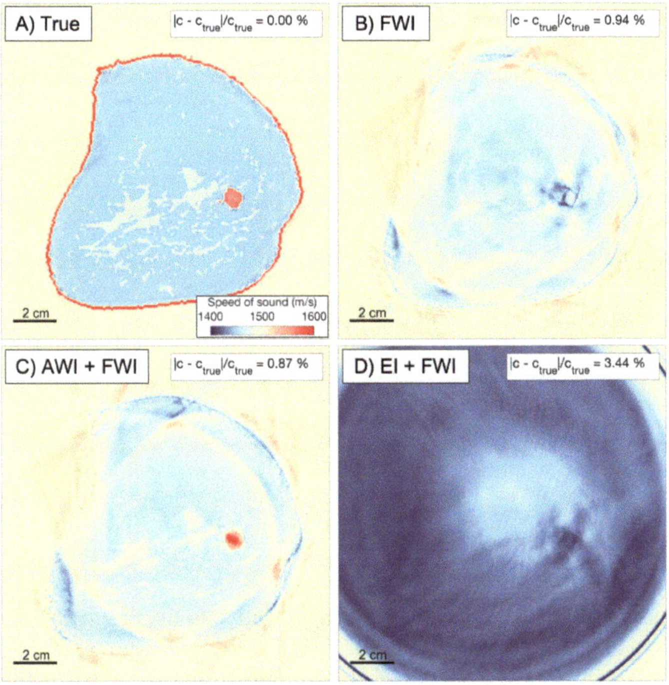

Figure 4: From A to D, true and recovered models obtained with FWI, AWI and EI both followed by FWI respectively starting from a homogeneous water starting model and a minimum frequency of $830 kHz$ for the modified USCT data challenge dataset.

is obtained in the laboratory by submerging the phantom in a water tank and scanning the target by moving two P4-1 linear ultrasound probes, transmitting at a central frequency of $1.4 MHz$, around the phantom in a 2D ring geometry.

Before inverting the data, we do some minimal data pre-processing. We first resample the data and extract a source wavelet from a water dataset that is acquired without the phantom. Additionally, we calibrate source positions using the watershot and a local minimisation method. The SNR of the data is good enough to start the inversion at $500 kHz$, and we progressively filter the data at increasing frequency bands up to $700 kHz$. Figure 5b displays the recovered speed of sound model obtained with FWI using a water starting model. We observe that FWI recovers some of the features of the inner layers, including the water-filled region in the middle of the phantom, but it is predominantly contaminated by cycle-skipping artefacts that blur the outline of the phantom, negatively affecting the resolution and accuracy of the recovered model. On the other hand, Figure 5c shows that the combination of AWI and FWI results in a much better resolved and accurate model with good recovery of the foldings of the outer PVA layer, the water-filled middle region, and the outline of the phantom. Nevertheless, there are inaccuracies in the upper part of the model which may be associated to incorrect calibration of source positions in this region, to which AWI is not immune.

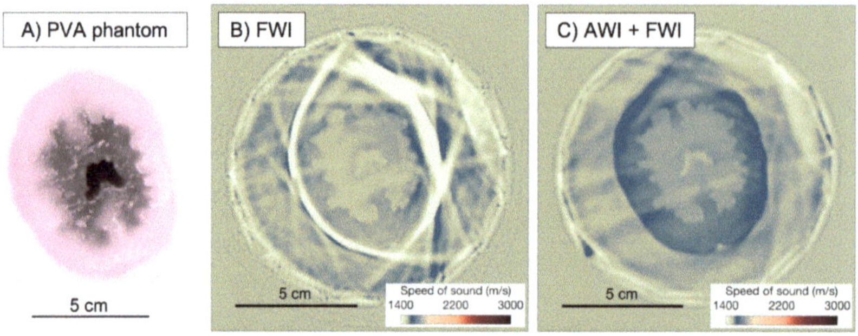

Figure 5: Ultrasound data are acquired in the laboratory for the soft PVA brain phantom in A) (photo) with two standard P4-1 probes transmitting at $1.4MHz$. The data are inverted utilising a homogeneous water starting model and using FWI and AWI followed by FWI, which respectively results in the speed of sound models in B) and C).

In summary, these results show that FWI is very sensitive to the absence of low frequencies when applied to a laboratory dataset, whereas AWI is more immune to cycle-skipping effects for soft tissue imaging. The next section evaluates their performance in silico in the presence of an outer hard-tissue layer.

3.3 In-silico brain and skull phantom data

We demonstrate here that FWI and AWI have a great potential for imaging the brain through the skull on an in-silico simulation. For this aim, we use a numerical phantom that mimics that in Figure 5a, but the model is scaled up so that it has the dimensions of a human head and an outer hard-tissue layer that surrounds the soft tissues, as displayed in Figure 6a. This layer has a significantly higher speed of sound of $3000m/s$. Ultrasound data was acquired by deploying a total of 128 transducers around the phantom in a circular geometry, which emit a three-cycle toneburst at a central frequency of $400kHz$. The use of such low frequencies ensures there is good penetration through the skull and little attenuation, while also minimising cycle-skipping effects.

The first test is designed to evaluate the potential of FWI in reconstructing the soft tissues in the interior of the phantom when the skull is known accurately. Hence, for this test we use the skull model in Figure 6b and perform 132 FWI iterations starting at a frequency of $100kHz$ up to $850kHz$, which results in the model in Figure 6c. All regions inside the phantom are accurately reconstructed with good resolution and speed of sound values that are very close to the true ones, including the water-filled region in the middle of the model. However, when prior information of the skull is unavailable and inversions start from a homogeneous water starting model in Figure 6d, FWI converges to the cycle-skipped model in Figure 6e. Given that the starting model is far from the true one, FWI gets trapped in a local minimum and fails in reconstructing the skull model and, consequently, the inner soft tissues. However, the recovered model after AWI followed by FWI in Figure 6f more accurately recovers the skull position and its speed of sound, which helps the inversion converge to a model that is closer to the true model. The three distinct soft tissue regions inside the phantom are well recovered, despite some artefacts caused during the initial iterations of AWI, in which the skull is still unknown.

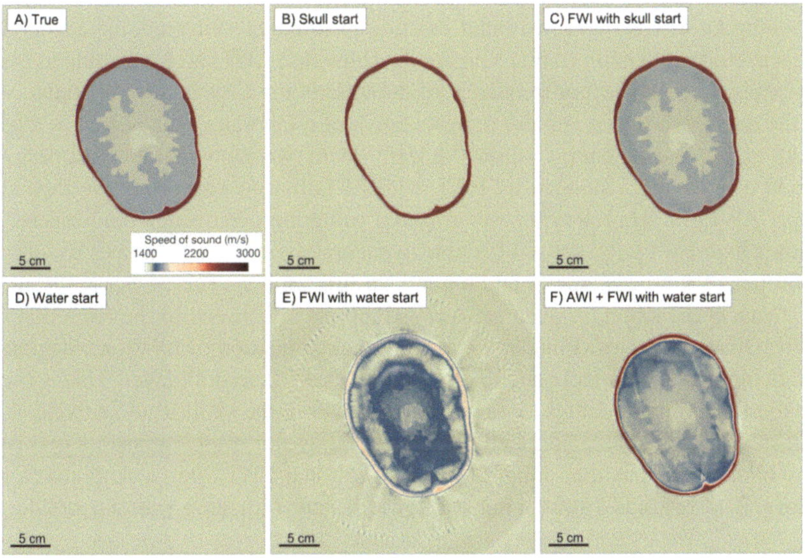

Figure 6: Ultrasound data are synthetically generated for the in-silico speed of sound phantom in A), which contains both soft and hard tissues simulating those of a human head. The data are inverted from $100 kHz$ to $850 kHz$ by first using the starting model in B) and FWI, resulting in the model in C). Then, the homogeneous water model (i.e. no prior skull information) in D) is used as a starting model for FWI, which results in the model in E), whereas F) shows the improvement obtained using AWI followed by FWI.

This in-silico test demonstrates the importance of using a functional that is immune to cycle-skipping to account for the generally unknown properties of the skull, and also shows that AWI combined with FWI has the potential to enable neuro-imaging with ultrasound. We therefore encourage further research into formulations and strategies that mitigate cycle-skipping effects, particularly for imaging through hard tissues. Finally, we note that it is straightforward to extend the 2D tests performed here to 3D, leading to similar conclusions.

4 Conclusions

The application of full-waveform inversion within medical imaging is still in its infancy, yet it has already shown great potential for soft-tissue applications, such as breast imaging. In this paper we address the problem of cycle-skipping - a key limitation of FWI - demonstrating that it can lead to incorrect recovered models in the absence of low frequency data or good starting models. To explore this, we first have successfully inverted the 2D blind dataset from the 2019 USCT data challenge. Using this dataset, we then tested the resilience to cycle-skipping of FWI and two other existing techniques - envelope inversion (EI) and adaptive waveform inversion (AWI). Our results show that AWI and EI are able to overcome cycle-skipping when a starting frequency of $500kHz$ is used, whereas FWI fails. We then modify the dataset so that it is more representative of a laboratory dataset, with a narrower bandwidth and no usable energy below $0.8MHz$. Here, we show that the best performance is obtained using AWI, seconded by FWI, with EI failing to converge. We then successfully apply AWI to a laboratory dataset acquired using high-frequency transducers. This is a situation where FWI fails, yet AWI is able to accurately image soft-tissues using real data in the absence of low-frequencies and a good starting model. This is especially important for breast cancer detection, as accurate reconstructions are crucial to the reduction of false positive/negative diagnoses. Finally, we explore the application of FWI for imaging within an in-silico human head model containing an outer hard-tissue skull layer. This is especially problematic for FWI as the high acoustic velocity within the skull leads to cycle-skipping, and skull starting models are not readily available. However, we demonstrate that AWI followed by FWI overcomes the limitations of conventional FWI when a skull starting model is unknown. This result is a substantial step towards neuro-imaging with ultrasound.

References

[1] M. A. Brown, R. C. Semelka, and B. M. Dale, *MRI: basic principles and applications.* John Wiley & Sons, 2015.

[2] W. A. Kalender, "X-ray computed tomography," *Physics in Medicine & Biology*, vol. 51, no. 13, p. R29, 2006.

[3] C. Li, N. Duric, P. Littrup, and L. Huang, "In vivo breast sound-speed imaging with ultrasound tomography," *Ultrasound in medicine & biology*, vol. 35, no. 10, pp. 1615–1628, 2009.

[4] P. Williamson, "A guide to the limits of resolution imposed by scattering in ray tomography," *Geophysics*, vol. 56, no. 2, pp. 202–207, 1991.

[5] Ò. Agudo, L. Guasch, P. Huthwaite, and M. Warner, "3D imaging of the breast using full-waveform inversion," in *Proceedings of the International Workshop on Medical Ultrasound Tomography*, pp. 99–110, 2018.

[6] M. Warner, A. Ratcliffe, T. Nangoo, J. Morgan, A. Umpleby, N. Shah, V. Vinje, I. Štekl, L. Guasch, C. Win, *et al.*, "Anisotropic 3D full-waveform inversion," *Geophysics*, vol. 78, no. 2, pp. R59–R80, 2013.
[7] M. Warner and L. Guasch, "Adaptive waveform inversion: Theory," *Geophysics*, vol. 81, no. 6, pp. R429–R445, 2016.
[8] R.-S. Wu, J. Luo, and B. Wu, "Seismic envelope inversion and modulation signal model," *Geophysics*, vol. 79, no. 3, pp. WA13–WA24, 2014.
[9] L. Métivier, A. Allain, R. Brossier, Q. Mérigot, E. Oudet, and J. Virieux, "Optimal transport for mitigating cycle skipping in full-waveform inversion: A graph-space transform approach," *Geophysics*, vol. 83, no. 5, pp. R515–R540, 2018.
[10] G. Yao, N. V. da Silva, M. Warner, D. Wu, and C. Yang, "Tackling cycle skipping in full-waveform inversion with intermediate data," *Geophysics*, vol. 84, no. 3, pp. R411–R427, 2019.

Time-Domain Full Waveform Inversion for High Resolution 3D Ultrasound Computed Tomography of the Breast

F. Lucka[1,2], M. Perez-Liva[3], B. Treeby[2], and B. Cox[2]

[1] *Centrum Wiskunde & Informatic, E-mail: felix.lucka@cwi.nl*
[2] *University College London, London, United Kingdom*
[3] *Hopital Européen Georges Pompidou, Centre de recherche Cardiovasculair, Paris, France*

Abstract

Ultrasound computed tomography (USCT) systems, which can image the breast without delivering harmful radiation, are typically designed to provide images of 2D slices, rather than full 3D images, to reduce the challenges of data acquisition and image reconstruction. Recently, however, novel scanners are being developed which allow the breast to be probed ultrasonically in three-dimensions. In particular combining photo-acoustic tomography and USCT in 3D promises access to high-quality images of tissue properties with important diagnostic value. However, the image reconstruction problems involved come with severe computational challenges.

In this talk, we focus on the problem of 3D speed-of-sound (SOS) imaging. The large size of the breast precludes using very high frequency US signals as they are damped too strongly. Time-of-flight reconstruction methods that rely on ray-based approaches are less accurate in the low-frequency regime and need very many data acquisitions as they rely on identifying single events in the recorded pressure-time series. Full Waveform Inversion (FWI) approaches are promising in this situation, because they do not assume ray-like propagation and can exploit the full time series data. However, they are challenging from multiple perspectives. Here, we demonstrate how to combine different computational techniques to obtain accurate 3D FWI results of the human breast with high spatial resolution. First, we rely on fast time-domain k-space pseudospectral methods implemented on GPUs for simulating the wave propagation. Second, we bypass the prohibitive memory requirement of standard adjoint-state-based gradient computation methods using a novel approach based on time-reversal. Finally, we combine numerical source encoding techniques with a stochastic variant of the L-BFGS method into a fast numerical optimization scheme.

High-Frequency Full-Waveform Inversion for Ultrasound Transmission Tomography

J. L. Herraiz[1], M. Perez-Liva[2], E. Mercep[3], X.L. Deán-Ben[4,5], D. Razansky[4,5], and J.M. Udias[1]

[1] *Complutense University of Madrid, Grupo de Fisica Nuclear and IPARCOS, Madrid, Spain*
E-mail: jlopezhe@ucm.es
[2] *Hôpital Européen Georges Pompidou, Centre de recherche Cardiovasculair, Paris, France*
[3] *iThera Medical GmbH, Munich, Germany*
[4] *University of Zurich, Zurich, Switzerland*
[5] *ETH Zurich, Zurich, Switzerland*

Abstract

Full-waveform inversion (FWI) algorithms are known to yield the best image quality for Speed-of-Sound (SoS) reconstruction in Ultrasound Transmission Tomography (USTT). Nevertheless, FWI solves a complex variational problem, and its memory and time requirements largely increase with the frequency of the US signal, making it impractical for many high-frequencies (¿3MHz) acquisitions. The spatial resolution required to correctly appreciate details in small animals force the use of high ultrasound (US) frequencies. For example, at least 5 MHz is needed to observe details of 0.5 mm or less in the reflectivity images. This has limited the massive application of FWI methods in the preclinical field.

In this work, we have introduced a methodology that allows overcoming two of the most important limitations of FWI methods: the stability of their convergence and the dependence of their computational cost on the frequency content of the signals. The first proposed modification was the generalization of the cost function (CF) to make the convergence more robust. In the initial iterations, the CF is dominated by a time-of-flight (TOF) misfit, which provides low-resolution images, but avoids convergence issues. As iterations progress, the CF becomes dominated by the standard FWI mean-square-error between the measured and the estimated data, which yields higher resolution images. The second modification was to simplify the wave propagation model. We obtained the estimated ultrasound field as the convolution of a reference waveform with the TOF information from many different paths between the emitter and the receiver. This approach is fast, flexible (it can be easily applied to 3-D data), and sufficiently accurate for most applications.

We implemented the proposed FWI algorithm using the large computing capabilities of current GPUs. It was evaluated with a tissue-mimicking gelatin phantom and a mouse acquired with a hybrid optoacoustic-ultrasound (TROPUS) device, which consists of a ring of 512 transducers operating at a central frequency of 5 MHz.

The SoS maps were reconstructed with this FWI method in about 1 minute/slice. The images obtained are quantitatively accurate and show significantly better resolution (around 1mm) compared to bent-rays methods. In conclusion, we proposed FWI algorithm that can be used to reconstruct SoS maps from real data acquired at high-frequencies, achieving high-quality images fast-enough for practical applications.

Analysis of linearized inverse problems in ultrasound transmission imaging

H. Wang[1], H. Gemmeke[2], K. W. A. van Dongen[3], T. Hopp[2], and J. Hesser[1]

[1] *Medical Faculty Mannheim, Heidelberg University, Mannheim, Germany, E-mail: hongjian.wang@medma.uni-heidelberg.de*
[2] *Institute for Data Processing and Electronics, Karlsruhe Institute of Technology, Karlsruhe, Germany*
[3] *Faculty of Applied Sciences, Delft University of Technology, Delft, Netherlands*

Abstract

The purpose of this paper is to analyze the linearized inverse problem during the iterative solution process of the ill-posed nonlinear inverse problem of image reconstruction for ultrasound transmission imaging. We show that the conjugate gradient applied to normal equation (CGNE) method gives more reliable solutions for linearized systems than Tikhonov regularization methods. The linearized systems are more sensitive when treated by CGNE than by Tikhonov regularization methods. The Tikhonov regularization is less effective at the beginning of the outer-loop iteration, where the nonlinearity is dominating while the conjugate gradient for the linearized system stops earlier. Only when the linear approximation is good enough to describe the whole system, Tikhonov regularization can fully play its role and give slightly better reconstruction results as compared to CGNE in a very noisy case.

Keywords: Gauss-Newton method, Inverse problem, Sensitivity analysis, Tikhonov regularization, USCT

1 Introduction

Ultrasound computed tomography (USCT) as imaging method offers high potential for breast cancer diagnosis. Due to the acquisition of transmission and reflection data from current 2D or 3D USCT devices [1-4], three types of images can be reconstructed from the acquired signals: reflection, attenuation and speed of sound (SoS) images [5]. While reflection images [6, 7] reveal changes in the echo texture of the surfaces between different tissues (qualitative imaging), transmission tomography offers quantitative characterization of the imaged tissue or materials by SoS and attenuation profiles [8].

The process of transmission imaging of the USCT system at *Karlsruhe Institute of Technology* (KIT) [1, 4] is separated into two parts: a forward model approximating the path of the ultrasound wave traveling through the 3D aperture; and an inverse problem given by the forward model and the measured signals to reconstruct the image. Our forward model is based on a paraxial approximation of the wave equation [9-11], which computes the pressure field $p(\eta)$ for known complex refraction index $1+\eta$. The inverse problem is hereby formulated as a nonlinear least-squares problem

$$f(\eta) = \frac{1}{2}\|p(\eta) - \hat{p}\|_2^2 = \frac{1}{2}\|r\|_2^2, \tag{1}$$

where $r : \mathbb{C}^n \to \mathbb{C}^m$ is called *residual vector*. Here, $\hat{p} \in \mathbb{C}^m$ are the pressure fields measured at the receivers, and $p(\eta) : \mathbb{C}^n \to \mathbb{C}^m$ are the predicted pressure fields computed according to the forward model (2). By minimizing $f(\eta)$ we want to reconstruct for the SoS and attenuation parameters that are inherent in η.

For the 3D USCT system reported in [1], a number of A-scans are performed to reconstruct a volume of voxels for a breast with a diameter of about 16 cm. As the validity of the paraxial approximation for forward angles is only up to 20°, the number of usable A-scans for transmission tomography is limited to about 6%. This is the main reason that the reconstruction using the paraxial approximation is an ill-posed problem. To solve this large-scale nonlinear inverse problem, we use a Newton type method yielding a set of linearizations of the problem (1). This is a standard approach for nonlinear inverse problems [12]. Due to the ill-posed nature of the problem, regularization is necessary. We choose three common methods of regularization for solving the linearized sub-problems in an inner loop: *conjugate gradient applied to normal equation* (CGNE) [13, 14]; damped least-squares [15, 16]; and gradient magnitude. The latter two use Tikhonov regularization.

In this paper, we specifically focus on analyzing the properties of the numerical solutions of the linearized inverse problems arising from Newton type iterations, with common regularization techniques. By this analysis we aim to show how well the predicted data match the true data and how close a particular estimate of the model parameters is to the true solution. For this type of problems, we propose using standard methods for analyzing linear inverse problems by computing the data resolution and model resolution [16] at different iterations. We choose standard Tikhonov type [15] regularization methods because they have closed-form solutions in these matrices.

2 Method

2.1 Forward model

The basis for transmission tomography is the wave propagation of ultrasound including refraction, diffraction, and forward scattering which is mathematically described by the wave equation for inhomogeneous media [17]. A full solution of the wave equation is computationally highly demanding and in practice an approximation is necessary. We make the assumption of a constant density fulfilled in the soft tissue of the breast, and use the paraxial approximation [9-11] where the forward solution of frequency dependent pressure field on the computational grid can be formulated as

$$p_{z+1} = e^{i\Delta z k_0 \eta_z} \mathscr{F}^{-1} \left\{ e^{i\Delta z \sqrt{k_0^2 - \xi^2}} \mathscr{F}\{p_z\} \right\}. \tag{2}$$

The acoustic medium is described by the background wave number $k_0 = \omega / c$ and the refractive index $1+\eta$, where $\omega = 2\pi f$ is the angular frequency for frequency f and spped of sound c of the background medium, and η accounts for the deviation of the inhomogeneity from the background medium. The forward solution is considered as a set of parallel slices perpendicular to the emission direction, where the index z for variables p and η denotes the considered z slice, whereas the indices for the (x,y)-directions are omitted. The spectral variable is denoted by ξ and the discrete Fourier transformation and its inverse are denoted by $\mathscr{F}$ and $\mathscr{F}^{-1}$ in 1D or 2D, depending on whether the problem is considered in 2D or 3D respectively.

2.2 Inversion by Gauss-Newton conjugate gradient

Image reconstruction estimates SoS, denoted as c, and attenuation, denoted as Δatt, which are incorporated in the complex variable η. To be specific, $\eta = a + i\frac{b}{\omega}$, where $\mathrm{Re}(\eta) = a = c_0/c - 1$ describes the deviation of c from the SoS c_0 in the background medium; and $\mathrm{Im}(\eta) = b/\omega = \Delta\mathrm{att}$ accounts for the deviation in the attenuation. We estimate η via solving the least-squares inverse problem (1) in a two-level strategy, by an outer and an inner loop. At each iteration of the outer loop, we linearize and reformulate the inverse problem using the Gauss-Newton (GN) method, which can be viewed as a modified Newton's method [18]. Specifically, given the nonlinear least-squares problem $f(\eta)$ of (1), instead of solving the standard Newton equations $\nabla^2 f(\eta) = -\nabla f(\eta)$ for a search direction d (which can be over-determined or under-determined depending on matrix dimensions), we solve the following system, i.e. the normal equations, to obtain the search direction

$$J^T J d = -J^T r. \tag{3}$$

Here, the derivatives of $f(\eta)$ are expressed in terms of the Jacobian J, which is the $m \times n$ matrix of the first partial derivatives of the residual vector, defined by

$$J = \left[\frac{\partial r_j}{\partial \eta_i}\right]_{\substack{j=1,2,\ldots,m \\ i=1,2,\ldots,n}} = \left[\nabla r_1^T, \nabla r_2^T, \ldots, \nabla r_m^T\right]^T. \tag{4}$$

The use of the approximation $\nabla^2 \approx J^T J$ relieves us to compute individual residual Hessians $\nabla^2 r_j, j = 1,2,\ldots,m$. To solve the linearized system (3), where the system matrix now corresponds to $J^T J$, we use the conjugate gradient (CG) method [19] as an inner loop solver.

2.3 Regularization

The minimization problem of the least-squares (1) is ill-posed, making regularization necessary. We analyze three methods for solving the linearized systems in the inner loop.

Method No.1: CGNE. The Gauss-Newton CG method applies CG to the normal equations of the linearized problem (3). As discussed in [13], this method has a regularization effect which comes from early termination of the iteration.

Method No.2: Damped least-squares regularization (denoted as Reg-DLS). A commonly used regularization method is *Tikhonov* regularization [15]. In our reconstruction problem, the goal is to estimate η in (2). Note that the GN search direction d in (3) is actually an update that will be added to the current guess of η at a given outer-loop iteration. If there exists an estimate of $\hat{\eta}$ that is close to the true η, then a Tikhonov regularization term can be defined to minimize the difference between $\hat{\eta}$ and the current guess. For example, let us consider η_{cur} as the best estimate so far for a given outer-loop iteration. We can simply assume $\hat{\eta} = \eta_{\text{cur}}$. Then, the regularization term for the linearized system at this outer-loop iteration is defined as

$$\|\eta_{\text{new}} - \hat{\eta}\|_2^2 = \|\Delta\eta\|_2^2 = \|\alpha d\|_2^2. \tag{5}$$

Accordingly, solving the linearized system (3) with this regularization term is equivalent to solving

$$\min_d \|Jd - r\|_2^2 + \lambda^2 \|d\|_2^2. \tag{6}$$

where λ is a regularization parameter. The minimizer solution d is obtained by solving

$$\left(J^T J + \lambda^2 I\right) d = J^T r. \tag{7}$$

Method No.3: Gradient magnitude regularization (denoted as Reg-GM). Another Tikhonov regularization term is based on a smoothness assumption using the discretized gradient as filter $[-1,1]$. Therefore, the regularization term is defined to penalize the gradient magnitude (GM) of η, i.e.

$$\|L\eta_{\text{new}}\|_2^2 = \|L(\eta + \alpha d)\|_2^2 \tag{8}$$

at a given outer-loop iteration, where

$$L = \begin{pmatrix} -1 & 1 & & & & \\ & -1 & 1 & & & \\ & & \dots & \dots & & \\ & & & & -1 & 1 \\ & & & & & 0 \end{pmatrix} \in \mathbb{R}^{n \times n}. \tag{9}$$

Accordingly, solving the linearized system (3) with this regularization term is equivalent to solving

$$\min_d \|Jd - r\|_2^2 + \lambda^2 \|L(\eta + d)\|_2^2. \tag{10}$$

The minimizer solution d is obtained by solving

$$\left(J^T J + \lambda^2 L^T L\right) d = J^T r - \lambda^2 L^T L\eta. \tag{11}$$

2.4 Analysis tools

We need to solve the linear inverse problem (3), or (7), or (11) at each outer-loop iteration of the reconstruction. We call d the *model parameters* and r the *data*, of the linear inverse problem $Jd = r$. Note that the term "model parameters" in the rest of this paper does NOT refer to the parameters η that we aim to reconstruct. We compute the *data resolution matrix* N and *model resolution matrix* R that are defined in [16], to respectively reflect how well the predicted data fits the observed data (via N), and how close a particular estimate of the model parameters is to the true solution (via R). In order to quantify the resolution quality, we use *Dirichlet spread functions* [16] which are based on the size, or *spread*, of the off-diagonal elements of resolution matrix. We also use the *Backus-Gilbert spread functions* [20], which is a weighted version of the spread functions.

3 Results

We compute the resolution matrices of the linearized problems arising from outer-loop iterations, when we apply different regularization methods and parameters. We simulate the measurements data of pressure field $\hat{p}$, based on the forward model of equation (2), as $p(\eta_{\text{exact}})$ plus additive Gaussian noise characterized by the signal-to-noise ratio (SNR). The known η_{exact} is the ground truth of a breast simulation. We only focus on 2D reconstruction problems, where the y dimension is omitted. The problem size n is the 2D phantom image size, i.e. $n = n_x \times n_z$, while the size of the measurements data depends on the number of transducers n_t, i.e. $m = n_x \times n_t$. For analysis purpose, at every outer-loop iteration, we have to compute, store and perform SVD or other operations on the system matrix of size $m \times n$, the data resolution matrix of size $m \times m$, and the model resolution matrix of size $n \times n$. The computational load and necessary memory are impractical for real-size problems. Therefore, we used a reduced problem setting as follows: the image size for all phantoms is $n_x \times n_z = 48 \times 36$ ($n = 1728$) with each pixel of size $\Delta z = 0.61$ mm; the radius of the measuring device is 18.2 mm and the radius of the phantom is 10.6 mm; $n_t = 128$ transducers are simulated at the frequency of 2.5 MHz. For the two Tikhonov methods Reg-DLS and Reg- GM, we set the regularization parameter λ by the L-curve method [15] applied at every outer-loop iteration. We perform an extra test for Reg-DLS with significantly larger λ values denoted as "Reg-DLS-lcurve2".

We plot the spread values of data resolution and model resolution matrices during 50 Gauss-Newton iterations for a noisy case using data with a SNR = 40dB, as shown in Figure 1. From our results, the data matrix N is a very sparse matrix, with large values located on the main diagonal and sub-diagonal positions. This means each row has a single sharp maximum centered about the main diagonal, and the data are well resolved. This phenomenon is testified by the upper chart in Figure 1, where the CGNE method has the smallest Dirichlet spread values when compared to any other Tikhonov regularization methods. This is because the Tikhonov regularization has the effect of preventing data over-fitting, where the predicted data are weighted average values of more neighboring observed data than the CGNE method, and hence are generally more biased. This property of the Tikhonov regularization results in more non-zero elements in each row of its data matrix N, and therefore larger Dirichlet spread values.

During the first few outer-loop iterations, the linearized systems are not stable, which is reflected by the fluctuations of the Dirichlet spread curves. As the reconstruction process continues, the spread values become stable after a certain number of outer-loop iterations, meaning that the results become more reliable. The R's Backus-Gilbert spread values from Tikhonov regularization methods also fluctuate during the first few outer-loop iterations. This indicates that the model parameters possess a bad ordering since the Backus-Gilbert spread favors a natural ordering in the model parameters, where the rows of R represent localized averaging functions [16]. However, for the parameters that we aim to reconstruct, they are mostly smooth and well-ordered, implying that the linearized approximations during

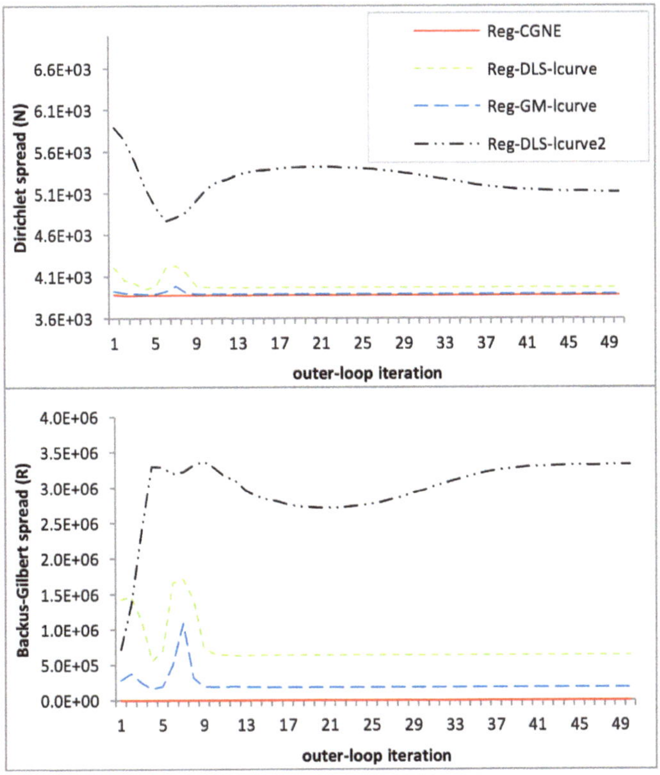

Figure 1: The spread values of data resolution and model resolution matrices during 50 Gauss-Newton iterations for noisy data with a SNR = 40 dB. The horizontal axis indicates the outer-loop iterations. **Top:** the Dirichlet spread of data resolution matrix. **Bottom:** the Backus-Gilbert spread of model resolution matrix.

the first few outer-loop iterations are not reliable. This is in agreement with our observations for the analysis of data resolution matrices.

We found that the method Reg-DLS-lcurve2 converges obviously slower than other methods. This method has the largest λ values, and therefore, the outer-loop iterative steps are more like gradient descent as compared to other methods whose steps are more like the Newton-type method. This explains the slower convergence. On the other hand, this method converges to slightly better reconstructions with our noisy data of SNR = 40dB.

4 Conclusions

We have analyzed the properties of the numerical solutions of the linearized inverse problems handled by common regularization techniques, arising from Gauss-Newton iterations in image reconstruction of ultrasound transmission tomography. To analyze the system sensitivity, we have computed the data resolution and model resolution of the linearized systems when applied to the CGNE method and to two Tikhonov regularization methods DLS and GM. Our analysis of the linear problems during the iterative solution process gives valuable information about the problem itself and yields good indications of the success of the solution process. Based on the analysis, a combination of different strategies, starting with CGNE and ending with Tikhonov would be reasonable.

References

[1] H. Gemmeke and N. V. Ruiter.: 3D Ultrasound Computer Tomography for medical imaging. NIM Nucl. Instr. & Meth. In Phys. Res. Vol. Ax (2007) 1057-65

[2] N. Duric, P. Littrup, P. Chandiwala-Mody, C. Li, S. Schmidt, L. Myc, O. Rama, L. Bey-Knight, J. Lupinacci, B. Ranger, A. Szczepanski, and E. West.: In-vivo imaging results with ultrasound tomography: Report on an ongoing study at the Karmanos Cancer Institute. Proc. SPIE vol. 7629 (2010)

[3] J. Wiskin et al.: Inverse Scattering and Refraction Corrected Reflection for Breast Cancer Imaging. Proc. SPIE vol. 7629 (2010)

[4] N. V. Ruiter, G. Göbel, L. Berger, M. Zapf, and H. Gemmeke.: Realization of an optimized 3D USCT. Proc. SPIE vol. 7968 (2011) 796805

[5] N. Ozmen, R. Dapp, M. Zapf, H. Gemmeke, N. V. Ruiter and K. W. van Dongen.: Comparing different ultrasound imaging methods for breast cancer detection. IEEE Transactions on Ultrasonics, Ferroelectrics, and Frequency Control 62(4) (2015) 637-646

[6] J. van der Neut, J. T. Fokkema, P. M. van den Berg, M. Zapf, N.V. Ruiter, U. Taskin, and K. W. A. van Dongen.: Ultrasonic synthetic-aperture interface imaging. IEEE Transactions on Ultrasonics, Ferroelectrics, and Frequency Control 66(5) (2019) 888- 897

[7] J. van der Neut, P. M. van den Berg, J. T. Fokkema and K. W. A. van Dongen.: Acoustic interface contrast imaging. Inverse Problems 34(11) (2018) 115006

[8] J. F. Greenleaf and R. C. Bahn.: Cinical imaging with transmissive ultrasonic computerized tomography. IEEE Transactions on Biomedical Engineering 28(2) (1981) 177-185

[9] H. Gemmeke et al.: Wave equation based transmission tomography. Proc. IEEE Int. Ultrason. Symp. (IUS) (2016) 1-4

[10] L. Althaus.: On acoustic tomography using paraxial approximations. M.S. thesis, Darmstadt University of Technology, Department of Mathematics, Darmstadt, Germany (2016)

[11] U. Taskin, N. Ozmen, H. Gemmeke and K. W. van Dongen.: Modeling breast ultrasound on the applicability of commonly made approximations. Archives of Acoustics 43(3) (2018) 425-435

[12] B. Kaltenbacher, A. Neubauer, and O. Scherzer.: Iterative Regularization Methods for Nonlinear Ill-Posed Problems. Radon Series on Computational and Applied Mathematics. De Gruyter (2008)

[13] M. Hanke.: Regularizing properties of a truncated newton-cg algorithm for nonlinear inverse problems. Numerical Functional Analysis and Optimization 18(9) (1997) 971-993

[14] M. Hanke.: Conjugate gradient type methods for ill-posed problems. Routledge (2017) Analysis of linearized inverse problems in ultrasound transmission imaging

[15] P. C. Hansen.: Regularization tools: A MATLAB package for analysis and solution of discrete ill-posed problems.: Numerical Algorithms 6(1) (1994) 1-35

[16] W. Menke.: Geophysical data analysis: Discrete inverse theory (MATLAB Edition). 3rd ed, Academic Press (2012)

[17] P. Suetens.: Fundamentals of medical imaging. 2nd ed. Cambridge University Press, Cambridge, UK (2009)

[18] J. Nocedal and S. Wright.: Numerical Optimization (Second Edition). Springer- Verlag New York (2006)

[19] J. R. Shewchuk.: An introduction to the conjugate gradient method without the agonizing pain. (1994)

[20] G. E. Backus and J. F. Gilbert.: Numerical application of a formalism for geophysical inverse problems. Geophys. J. Roy. Astron. Soc 13 (1967) 247-276

Time-of-Flight Picking for Ultrasound Computed Tomography of the Breast

F. Lucka[1,2] and B. Cox[2]

[1] *Centrum Wiskunde & Informatic, E-mail: felix.lucka@cwi.nl*
[2] *University College London, London, United Kingdom*

Abstract

Ultrasound Computed Tomography (USCT) is a promising imaging modality for detection of tumors in the breast. There are several approaches to reconstructing images from USCT data. Techniques that use the Time-of-Flight (TOF) of the ultrasound pulses across the breast are popular because of their computational efficiency, and historically because of their connection to the Radon transform, for a long time the workhorse of tomography. They have also been used to generate a starting point for more computationally-intensive, but potentially more accurate, approaches based on full-wave inversion. One key aspect of TOF methods is the accurate extraction of the TOF from the measured ultrasound pulses. Methods using cross-correlation perform well for homogeneous or low-scattering media, but their performance can dramatically deteriorate for more complex media because of the delay spread, or coda, that appears in the measured signal. Therefore, techniques using the first arrival time of the signals are often used. Typically, the performance of these algorithms is contingent on the shape and duration of the excitation pulse, the scattering properties of the intervening medium, and the signal-to-noise ratio (SNR) of the received signals. Despite the existence of many first-arrival-picking algorithms for the geophysical applications, finding an accurate arrival-time picking algorithm that performs well for all measured time traces remains a challenge. To cope with this limitation, picking the first arrival in an individual time trace is often followed by pragmatic techniques for detection and recalculation of mis-picks using correlation information from the entire measured data.

In this study, we exploit the additional information that exists for medical USCT, as opposed to geophysical applications, in order to apply additional constraints on the travel time picking. The information includes, but is not limited to the known excitation pulses, loose bounds on the maximal and minimal sound speed of the medium, the position of the transducers, and the limited number of large-amplitude peaks in the envelope of the time traces. Several time-of-flight picking algorithms are demonstrated using numerical simulations on a digital breast phantom obtained from MRI images. We show the performance of the proposed strategies using realistic excitation pulses with different pulse durations and different ranges of SNR in the measured signals.

Regularization by Registration: Utilizing Prior Knowledge to Accelerate Ultrasound Full-Waveform Inversion

C. Boehm[1], N. Korta Martiartu[1], and A. Fichtner[1]

[1] *ETH Zurich, Zurich, Switzerland, E-mail: christian.boehm@erdw.ethz.ch*

Abstract

Ultrasound computed tomography (USCT) is an emerging imaging modality to characterize human soft tissue, in particular for the application of breast cancer screenings. Typical USCT apertures collect both transmission and reflection data from various transducers acting either as emitters or receivers in a fixed acquisition geometry. Full-waveform inversion exploits the complete waveforms provided by those data to provide high-resolution quantitative images of several parameters, such as speed of sound, density, or attenuation.

Despite recent advances in hardware and algorithmic developments, time-domain acoustic waveform inversion remains computationally challenging, in particular for 3-D reconstructions and high-frequency data.

In contrast, conventional ray-based methods like time-of-flight inversion or B-mode imaging may suffer from limited resolution due to the simplified wave physics, but they are considerably cheaper to compute and readily utilize different parts of the measured signals.

In this contribution, we present strategies to utilize prior information on the model space generated from such auxiliary imaging problems to accelerate 3-D full-waveform inversion. This includes (1) construction of quantitative prior models based on time-of-flight inversions, (2) dimension reduction of the model space using shapes identified by B-mode imaging, and (3) reparameterization of the model space by using sparsifying constraints from segmentation. This is tied into a trust-region inversion framework that features a stochastic quasi-Newton method, encoded sources, total variation regularization, and spectral-element waveform simulations.

Several examples using numerical phantoms demonstrate the applicability of our approach to ring-shaped and fully 3-D acquisition geometries.

A Preclinical Simulation Study of Ultrasound Tomography for Pulmonary Bedside Monitoring

J. L. Mueller[1], D. Cárdenas[2], and S. Furuie[2]

[1] *Department of Mathematics and School of Biomedical Engineering, Colorado State University, Fort Collins, CO USA, E-mail: mueller@math.colostate.edu*
[2] *Escola Politécnica de Universidade de São Paulo, Brazil*

Abstract

Mechanical ventilation in the intensive care unit (ICU) is a life-saving technique for patients experiencing acute respiratory failure, but it is also associated with a high incidence of complications. Currently, there is no widely-used monitoring technique to guide the setting of the ventilator to facilitate a precision-medicine approach or to provide a real-time alert for developing adverse pulmonary conditions. Conventional ultrasound has been studied as a thoracic bedside technology for diagnosis and monitoring, but due to the lack of signal penetration into lung tissue, ultrasound images often contain more information in their artifacts than in the images themselves. However, low frequency (10 – 750 kHz) ultrasound has been shown to penetrate the lung, motivating its use as a non-ionizing tomographic technique for pulmonary monitoring. The rich acoustic data provided by the tomographic geometry, yields images that are easier to interpret and of higher resolution than conventional ultrasound and will enable continuous imaging by removing the need for a highly skilled technician. In this work, tomographic reconstructions of sound speed from numerically simulated low frequency ultrasound data on a belt of transducers on chest phantoms simulating three types of pneumothorax are compared. The distorted Born iterative method (DBIM) is used to compute the reconstructions in a K-wave implementation, and regularized reconstructions using Tikhonov regularization, the L^2 norm of the Laplacian, and total variation (TV) regularization are compared. We also demonstrate that excitation patterns in which all transducers both send and receive result in higher SNR and improved reconstructions over single-transmission patterns. These results support the feasibility of USCT as a non-ionizing bedside pulmonary monitoring technique for patients in the ICU, and the ability to identify a pneumothorax and its location and size.

Keywords: Ultrasound tomography, pulmonary imaging, low frequency

1 Introduction

Low frequency (LF) ultrasound has established medical applications in transdermal drug delivery [1], dentistry, eye surgery, kidney stone treatment, and blood clot elimination [2], and emerging medical-related applications include mapping the human vocal tract [3, 4] and mouth state detection [2]. Conventional ultrasound has been studied as a thoracic bedside technology for diagnosis and monitoring, but due to the lack of signal penetration into lung tissue, ultrasound images often contain more information in their artifacts than in the images themselves [5]. However, low frequency (10 – 750 kHz) ultrasound has been shown to penetrate the lung [6], and monitoring of air and water content in the lungs in this frequency range is feasible. In [7] LF ultrasound was used to detect air trapping in COPD patients. In [6] it is demonstrated that acoustic transmission between 1 Hz and 1 MHz through the thorax falls into three distinct regions: below 1 kHz the sound speed through the chest and lungs is between 30 and 50 m/s, between 1 and 10 kHz transmission is absent, and between 10 kHz and 1 MHz, acoustic transmission averages 1500 m/s.

The results in [6] motivate our study of the potential use of low-frequency ultrasound as a non-ionizing tomographic technique for pulmonary monitoring. The rich acoustic data provided by the tomographic geometry yields images that are easier to interpret and of higher resolution than conventional ultrasound and could enable continuous imaging by removing the need for a highly skilled technician. In this work, tomographic reconstructions of sound speed from numerically simulated LF ultrasound data on a belt of transducers on chest phantoms simulating three types of pneumothorax are compared. The distorted Born iterative method (DBIM) is used to compute the reconstructions in a K-wave [8] implementation, and regularized reconstructions using Tikhonov regularization, the L^2 norm of the Laplacian, and total variation (TV) regularization are compared. We also demonstrate that excitation patterns in which all transducers both send and receive result in higher SNR and improved reconstructions over single-transmission patterns. These results support the feasibility of USCT as a non-ionizing bedside pulmonary monitoring technique for patients in the ICU, and the ability to identify a pneumothorax and its location and size.

2 Distorted Born Iterative Method

The sound speed is reconstructed using the distorted Born iterative method (DBIM) [9, 10], a regularized Newton-based iterative method, which is summarized here with a brief derivation for completeness and clarity. The propagation of a time harmonic acoustic wave in an inhomogeneous medium Ω is modeled by

$$\rho(x)\nabla\cdot(\rho^{-1}(x)\nabla p(x)) + k^2(x)p(x) = -\phi^{inc}(x), \quad x \in \Omega \tag{1}$$

where $p(x)$ is the acoustic pressure, $\phi^{inc}(x)$ is the acoustic source, and $k(x) = \omega/c(x)$ is the wave number, where ω is the angular frequency of the incident harmonic wave, and $c(x)$ is the speed of sound in the medium. Assuming constant density, (1) becomes the Helmholtz equation

$$\Delta p(x) + k^2(x)p(x) = -\phi^{inc}(x), \quad x \in \Omega. \tag{2}$$

Consider a (nonhomogeneous) reference medium with wave number $k_0(x)$. Then the acoustic pressure $p_0(x)$ satisfies

$$\Delta p_0(x) + k_0^2(x)p_0(x) = -\phi^{inc}(x), \quad x \in \Omega. \tag{3}$$

Subtracting equation (3) from (2) results in

$$\Delta(p(x) - p_0(x)) + k_0^2(x)(p(x) - p_0(x)) = -q(x)p(x), \tag{4}$$

where

$$q(x) \equiv k^2(x) - k_0^2(x).$$

In the DBIM, the right-hand side of equation (4) is linearized by replacing p by p_0,

$$\Delta(p(x) - p_0(x)) + k_0^2(x)(p(x) - p_0(x)) = -q(x)p_0(x). \tag{5}$$

The solution to (5) can be written in terms of the Green's function $G_0(x)$ for the reference medium, which satisfies

$$\Delta G_0(x) + k_0^2(x)G_0(x) = -\delta(x). \tag{6}$$

The solution to (5) is given by

$$p(x) = p_0(x) + \int_\Omega G_0(x-y)q(y)p_0(y)dy. \tag{7}$$

Note for a homogenous reference medium in R^3 with wavenumber k_0, $G_0(r,r') = \frac{e^{ik_0|r-r'|}}{4\pi|r-r'|}$. However, the reference medium, and hence G_0 is updated each time a new approximation to $k(x)$ is computed. Denoting the pressures that are measured on the transducers by p_{meas}, the most recent approximation to $k(x)$ by $k^{(n)}(x)$, the corresponding Green's function satisfying (6) with k_0 replaced by $k^{(n)}(x)$, and the simulated pressures on the transducers corresponding to $k^{(n)}$ by $p^{(n)}(x)$, the cost functional to be minimized is defined by

$$F(q) = p^{(n)} + G_0 \star (qp^{(n)}),$$

where $q(x) = k(x) - k^{(n)}(x)$. Then we iteratively solve

$$\min_k \{ \|F(q) - p_{meas}\|_2^2 + \alpha L(q) \}$$

where $L(q)$ is a regularization term such as the L^2 norm $\|q(x)\|_2^2$, as in Tikhonov regularization, and α is the regularization parameter.

In this work we considered three different regularization terms: Tikhonov regularization, the L^2 norm of the Laplacian of q, and total variation (TV) regularization. The L^2 norm of the

Laplacian was implemented with a second order finite difference matrix. TV regularization was implemented by solving the minimization problem

$$\min_k \left\{ \|F(q) - p_{meas}\|_2^2 + \alpha \sum_{i=1}^{n} |L(q)_j| \right\}, \tag{8}$$

where L is the finite-difference matrix, using a steepest descent method. Since the gradient of the absolute value function is undefined, in the implementation the absolute value function is replaced by

$$|t|_\beta = \sqrt{t^2 + \beta}.$$

3 Methods

3.1 Data Simulation

Three phantoms simulating a cross-section of a human chest containing three sizes and locations of pneumothoraces in the left lung were constructed with sound speeds, density, and attenuation given in Table 1, where the attenuation is modeled in K-wave to follow a frequency power law [11], which allows efficient computation with Fourier spectral methods. The organ properties were the same in each of the three phantoms. The properties of the background medium were set to those of water. The torso was given a slightly higher sound speed and attenuation. Based on [6], the fluid matrix of the lung, comprised of blood and cellular matter, is the main mechanism of transport of the acoustic wave through the lung. For this reason, the healthy lung is assigned a sound speed of 1440 m/s and a higher attenuation than that of the torso. We recognize that this sound speed may be too high to represent that of the lung, and this will be an area of further experimental investigation. Since a pneumothorax contains little blood or cellular structure, the sound speed was modeled as lower than that of healthy lung, and the attenuation was modeled to be higher. All organs were modeled to have constant density to be consistent with the governing equations of our model. Plots of the sound speed distributions for each phantom are found in Figure 1.

	Tissue properties		
Simulated organ	Sound speed (m/s)	Attenuation $(dB/MHz^y cm)$	Density (kg/m^3)
medium	1540	0.0022	1000
torso	1550	0.01	1000
lungs	1440	0.8	1000
pneumothorax	1360	1.2	1000
spine	1640	2.0	1000
heart	1580	0.2	1000

Table 1: Sound speed, density, and attenuation in each organ for each of the three simulated pneumothoraces.

Simulated pressure data on 32 transducers were computed on a 3D mesh modeled as uniform in the z-direction in K-wave with a mesh size of 236 by 236 by 16. The transduc-

Figure 1: Sound speed distributions for the three simulated pneumothoraces. The colorbar scale is in m/s.

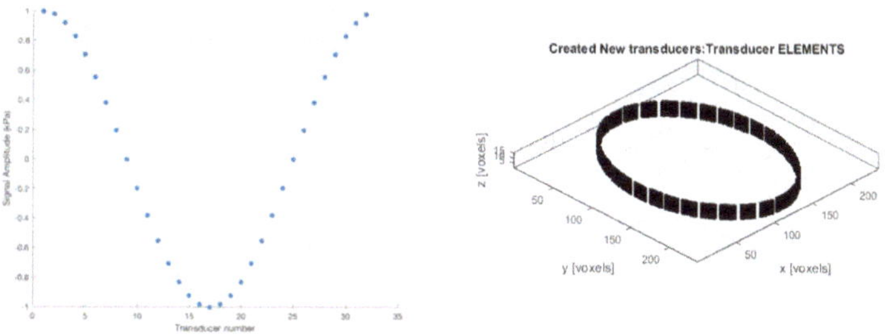

Figure 2: Left: Pressure amplitude on each of the 32 transducers for the first trigonometric excitation pattern. A negative sign simply inverts the signal. Right: Transducer positions in the computational model.

ers were evenly spaced around the boundary of the thorax as seen in Figure 2. A perfectly matched layer was included in the K-wave simulation to avoid reflections from the boundary. The incident signals were 5 cycles of a Gaussian pulse with center frequency 150 kHz and peak amplitude of 1000 Pa. Two types of incident transmission patterns were imposed for comparison. In [12] it is shown that excitation patterns that correspond to the eigenfunctions of the scattering map maximize the far-field energy, and hence the signal-to-noise ratio of the measured pressures on the transducers. For this reason, it is worthwhile to investigate patterns in which all transducers transmit and measure. Motivated by analogous results in electrical impedance tomography (EIT) [13], we define patterns in which all transducers transmit simultaneously, and the amplitude on each transducer is modulated according to its assigned number. Note that each transmission pattern must be linearly independent from the others to avoid redundancy in the data. As in EIT, the trigonometric functions $\{\cos(n\theta)\}$ and $\{\sin(n\theta)\}$ provide a natural choice of linearly independent functions to govern the amplitude of the signal on each transducer. For an even number L of transducers, we impose the $L-1$ linearly independent patterns $\{\cos(\theta), \cos(2\theta), \ldots, \cos(L/2\theta), \sin(\theta), \sin(2\theta), \ldots, \sin((L/2-1)\theta)\}$, where each choice of $\cos(n\theta)$ or $\sin(n\theta)$ defines a transmission pattern. Choosing one transducer to correspond to angle $\theta = 0$ as a reference, the amplitude of the signal on each transducer is then defined by its angle θ from the reference transducer according to the trigonometric function for that particular pattern. An example of signal amplitudes is provided in Figure 2.

A time snapshot of the pressure wave distribution arising from the first trigonometric excitation pattern for the small pneumothorax simulation is shown in Figure 3. Gaussian white noise with amplitude of 1% of the maximum received pressure over all transducers was added to the signals.

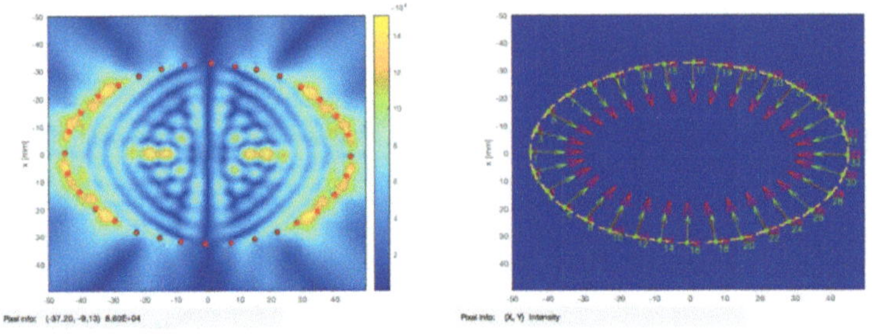

Figure 3: Left: Time snapshot of the pressure wave distribution arising from the first trigonometric excitation pattern for the small pneumothorax simulation. The colorbar scale is in Pa. Right: Vectors of normals to the transducers at each transducer center. The green arrow is the normal for the simulated data and red is for the reconstruction.

3.2 Solution of the Inverse Problem

The intermediate forward problem solutions were also computed in K-wave, but to avoid an inverse crime, a coarser mesh was chosen of size 220 by 220 by 16. Since the fine mesh is not a refinement of the coarse mesh, this resulted in small perturbations of the transducer positions and angles. The normal vectors to the center of each of the 32 transducers is shown in Figure 3 for the simulated data versus the reconstruction mesh.

There are two options for the updating step of the reconstruction algorithm. The first is to update the sound speed estimate after each transmission. In this case the Green's function is also updated after each transmission, with the source in the center, and equation (7) is computed with the current sound speed to obtain a new estimate for $p(x)$, the estimated pressure on the transducers. The second option is to update the sound speed estimate after all transmissions in the forward problem simulation have been completed, which we will refer to as a sweep.

4 Results and Discussion

Results of the reconstructions for the three simulated pneumothoraces are found in Figures 6 – 8 for trigonometric and single transmission excitation patterns. Results from Tikhonov regularization were consistently inferior to the other two methods and are shown only for the case of the large pneumothorax in the upper lobe in Figure 5. Plots of the root mean square error (RMSE) in each organ and throughout the torso are included, and the legend is reproduced in Figure 4 for readability.

It was found that for trigonometric excitation patterns, updating the Green's function after each transmission resulted in better images, while in the case of the single transmission excitation patterns, the method would diverge if the Green's function and sound speed estimate were updated after each transmission, but would converge if they were updated after each complete set of transmissions (each sweep). We conjecture that this difference is due to the richer set of information and higher SNR found in a single pattern of trigonometric excitation.

In all cases, a trade-off existed between the amount of oscillation, or speckle on the order of 3 to 5 mm, and the amount of regularization applied. In the reconstructed images TV regularization resulted in the least oscillation in the interior, the lowest RMSE's, and the most well-defined heart region. However, the speckle near the boundary is much higher for the TV regularization than for Tikhonov regularization or a Laplacian operator penalty term. While the overall trend of the logarithm of the objective functional and the RRE decrease with iteration number, both are oscillatory in the case of trigonometric excitation patterns, in which the Green's function and estimated pressures were updated with each transmission.

Ultrasound safety requirements are prescribed in terms of the mechanical index (MI) and thermal index (TI). The MI, introduced in [14], quantifies the potential for adverse biological effects caused by the occurrence of transient cavitation. It is defined by

$$MI = P_r/\sqrt{f}, \tag{9}$$

where P_r is the peak rarefractional pressure in vivo in MPa and f is the frequency in MHz. The BMUS recommendation is that the MI should be below 0.3 and the exposure time should be restricted for the lungs and intestines [15]. The US FDA recommends a MI less than 1.9 for all medical applications except opthalmology, for which it is lower [16]. In [17], it is observed that the extension of this definition of the MI to low frequencies is questionable, and an alternative formulation is presented and applied to cavitation in blood and in water. Referring to Figure 3, the maximum rarefactional pressure for the first trigonometric excitation pattern is 0.15 MPa, resulting in a MI of 0.38 for 150 kHz excitation. The alternative formulation indicates that a MI of 0.2 may be a better choice as an upper limit to avoid adverse effects from cavitation.

5 Conclusion

The study suggests that the use of ultrasound computed tomography with low frequency transmitted waves may be a feasible method for bedside detection of pneumothorax. Reconstructions of three simulated pneumothoraces were compared using the distorted Born

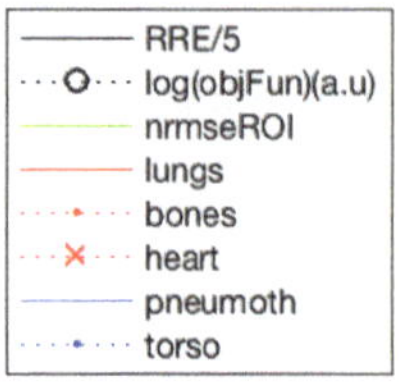

Figure 4: Legend for the RMSE in Figures 5 – 8

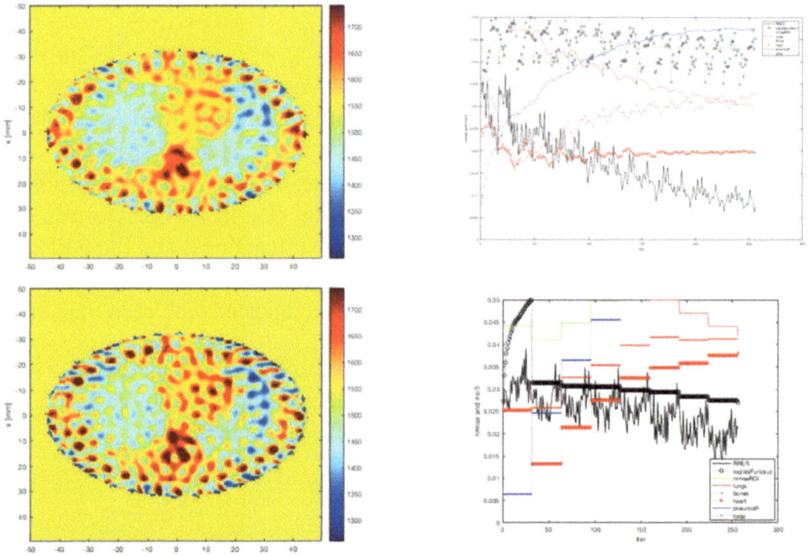

Figure 5: Left column: Reconstructions after 8 iterations (full sweeps) of the phantom with a large pneumothorax in the upper lobe using Tikhonov regularization. Right column: RMSE by organ. Top row: Results from trigonometric excitation, $\alpha = 0.01$. Bottom row: Results from single transmission excitation, $\alpha = 0.1$.

iterative method with three types of regularization and two types of acoustic excitation patterns. It was found that trigonometric excitation patterns – patterns in which all transducers are active simultaneously – resulted in more accurate reconstructions overall and higher distinguishability of the pneumothoraces. This suggests that the more complicated acoustic transmission patterns, which also result in a higher SNR, should be further investigated in an experimental setting for this application. Comparisons of the three methods of regularization suggest that total variation is the best.

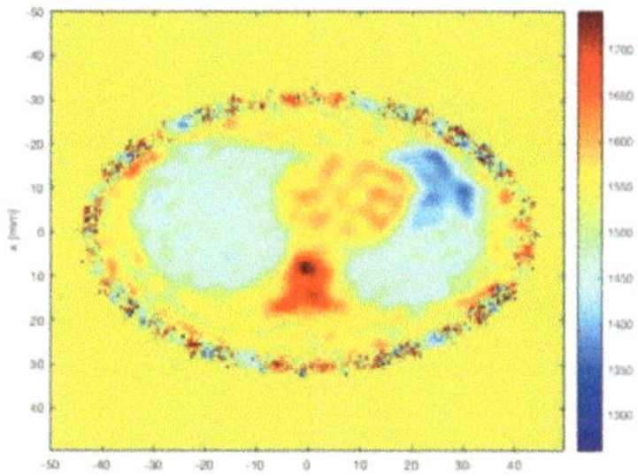

(a) Trigonometric excitation, TV $\alpha = 0.1$

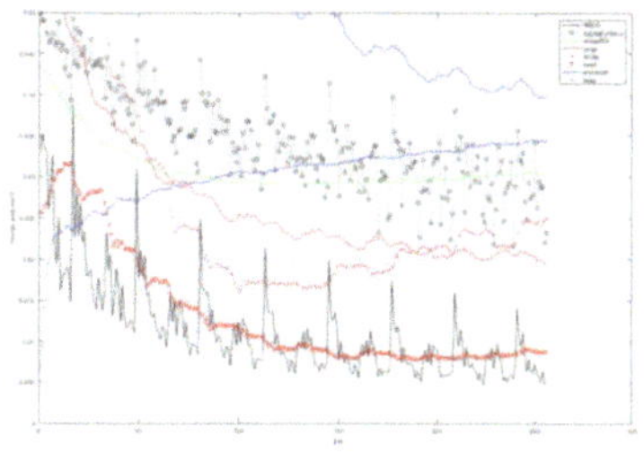

(b) Trigonometric excitation, TV, RMSE

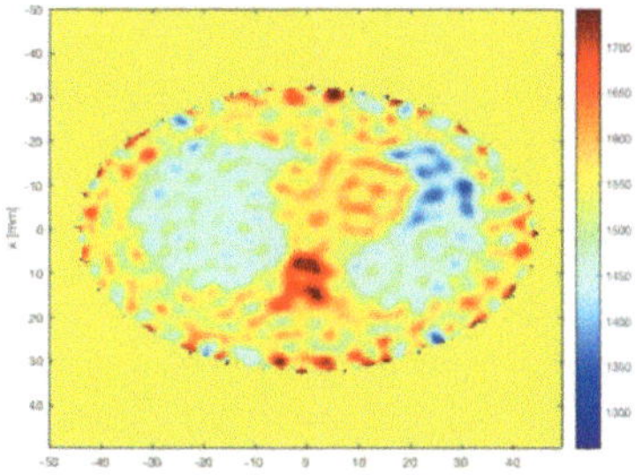

(c) Trigonometric excitation, Norm2Lapl $\alpha = 0.05$

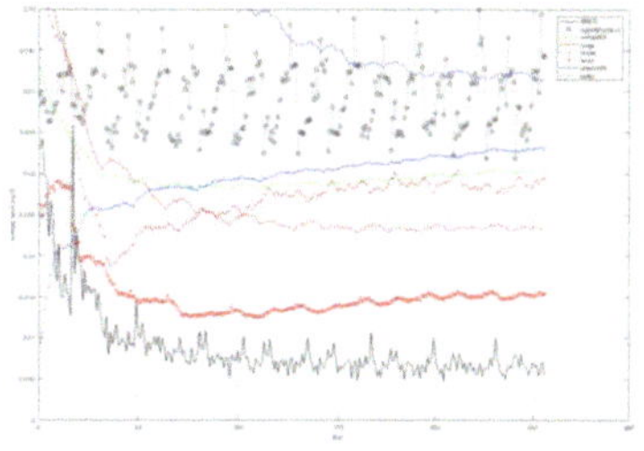

(d) Trigonometric excitation, Norm2Lapl, RMSE

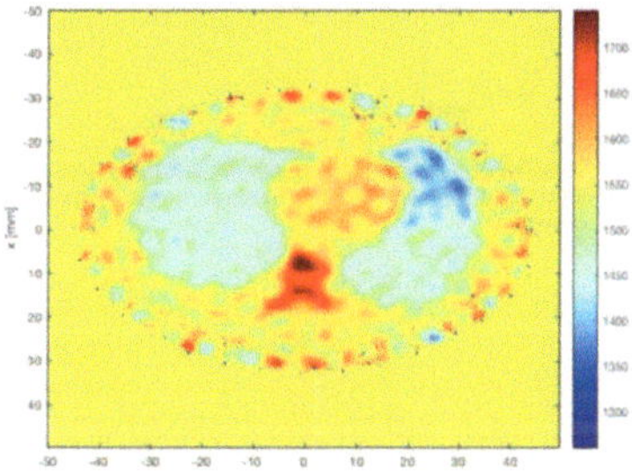

(e) Single transmission, TV $\alpha = 0.005$

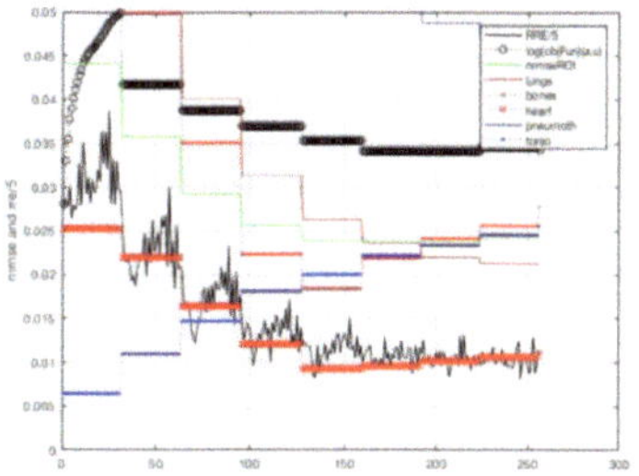

(f) Single transmission, TV, RMSE

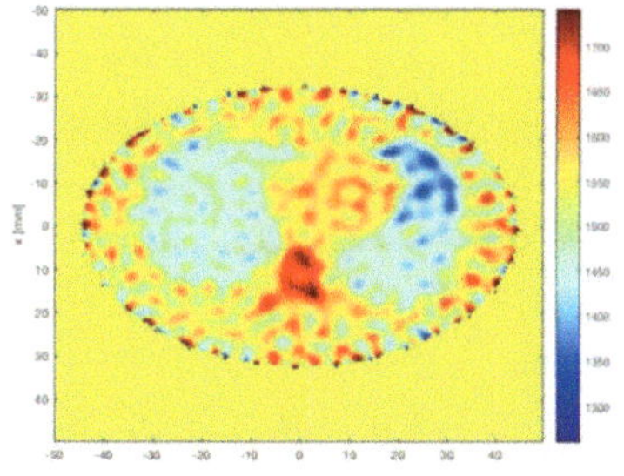

(g) Single transmission, Norm2Lapl $\alpha = 0.5$

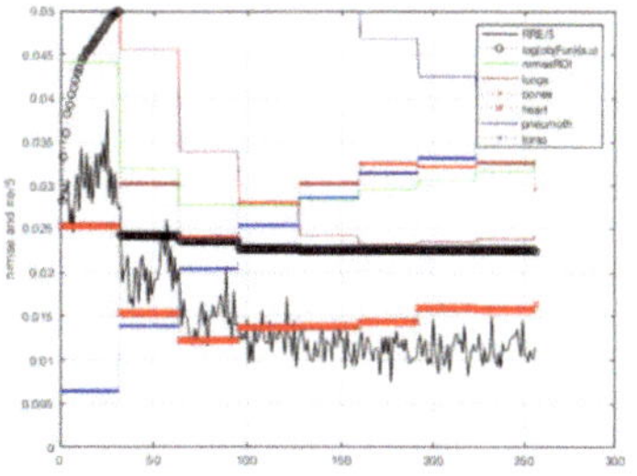

(h) Single transmission, Norm2Lapl, RMSE

Figure 6: Left column: Reconstructions after 8 iterations (full sweeps) of the phantom with a large pneumothorax in the upper lobe. Right column: RMSE by organ for each case.

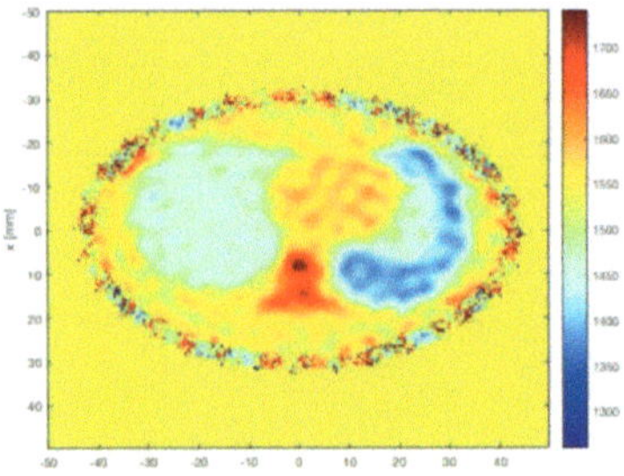

(a) Trigonometric excitation, TV $\alpha = 0.1$

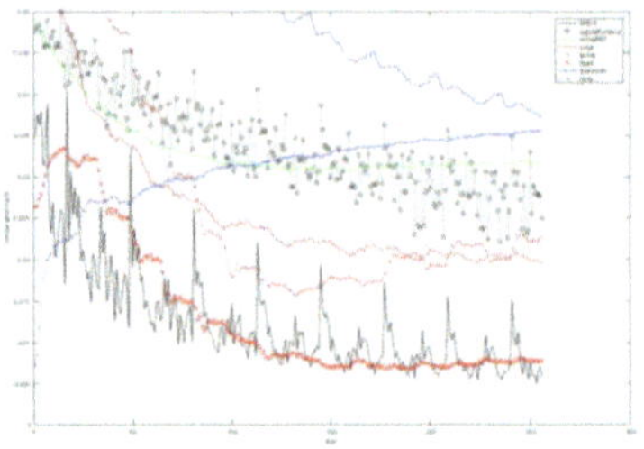

(b) Trigonometric excitation, TV, RMSE

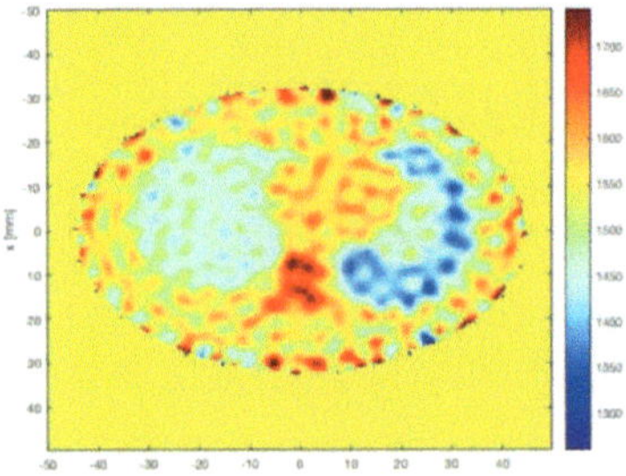

(c) Trigonometric excitation, Norm2Lapl $\alpha = 0.05$

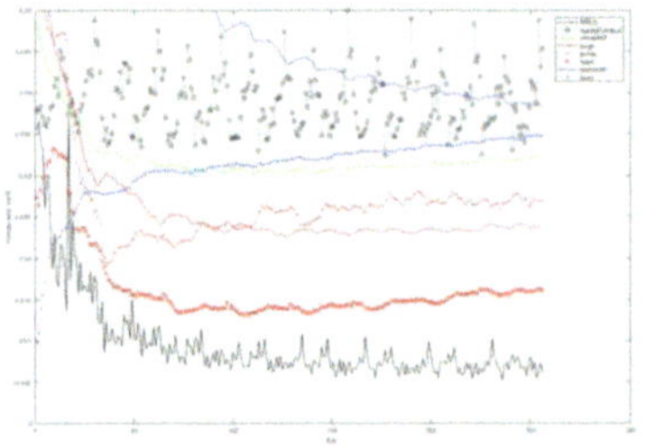

(d) Trigonometric excitation, Norm2Lapl, RMSE

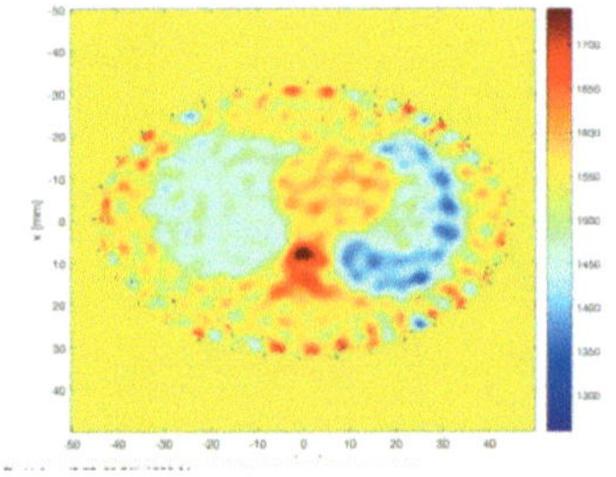

(e) Single transmission, TV $\alpha = 0.005$

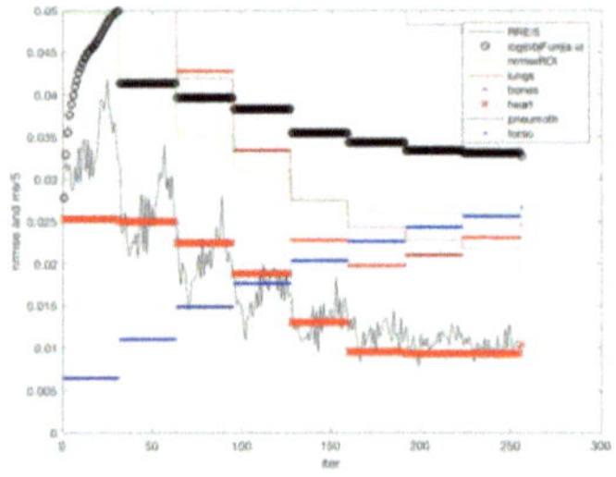

(f) Single transmission, TV, RMSE

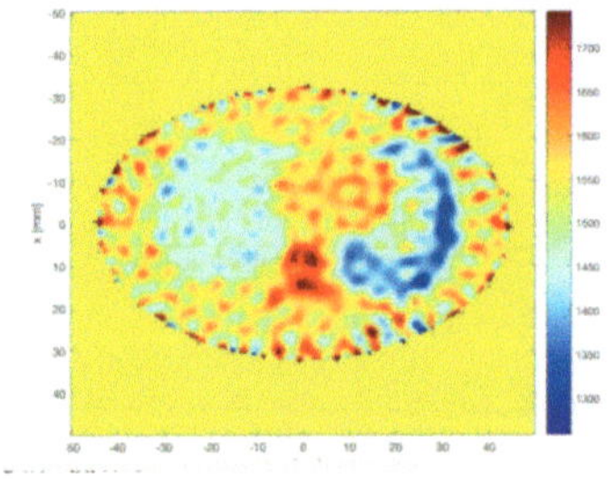

(g) Single transmission, Norm2Lapl $\alpha = 0.5$

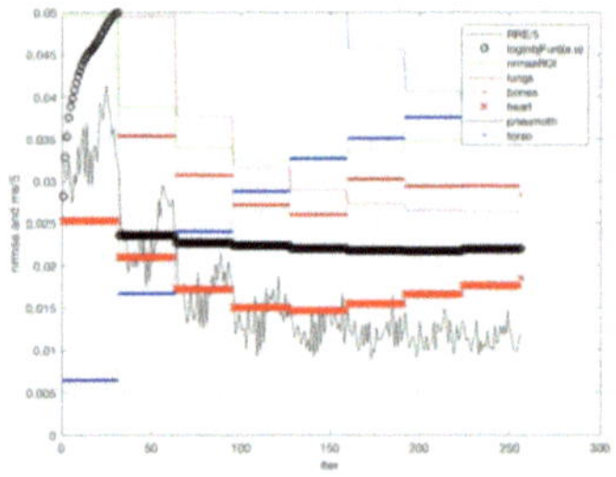

(h) Single transmission, Norm2Lapl, RMSE

Figure 7: Left column: Reconstructions after 8 iterations (full sweeps) of the phantom with a large pneumothorax in the left lung. Right column: RMSE by organ for each case.

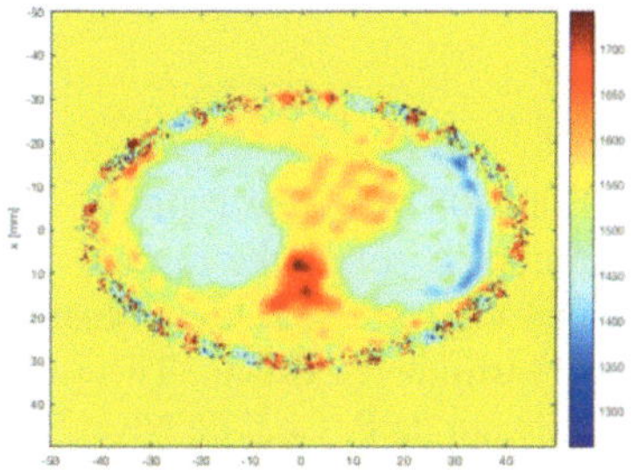

(a) Trigonometric excitation, TV $\alpha = 0.1$

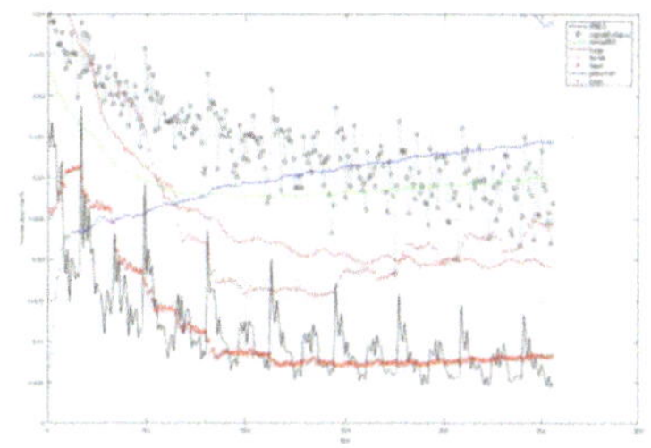

(b) Trigonometric excitation, TV, RMSE

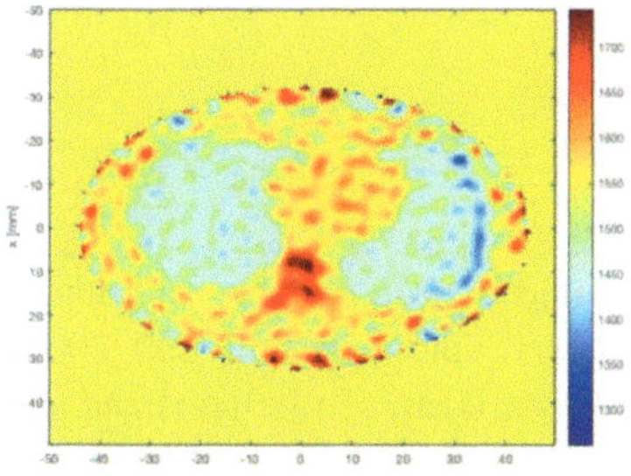

(c) Trigonometric excitation, Norm2Lapl $\alpha = 0.05$

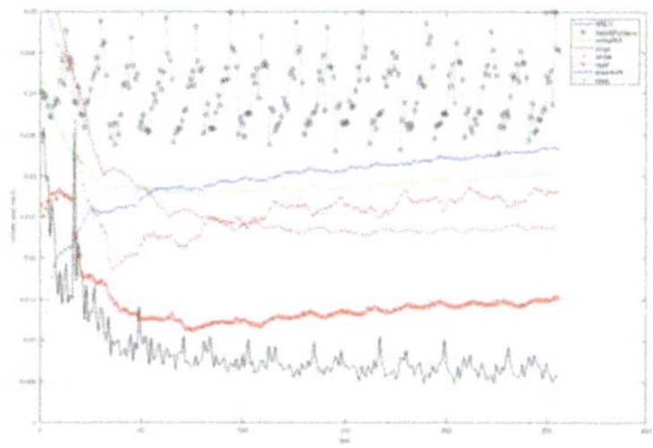

(d) Trigonometric excitation, Norm2Lapl, RMSE

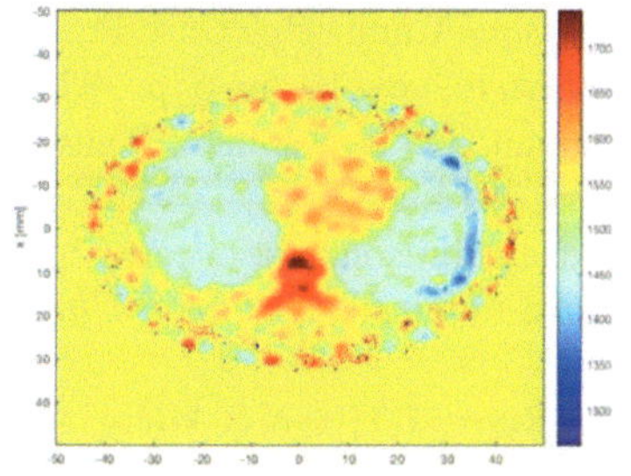

(e) Single transmission, TV $\alpha = 0.005$

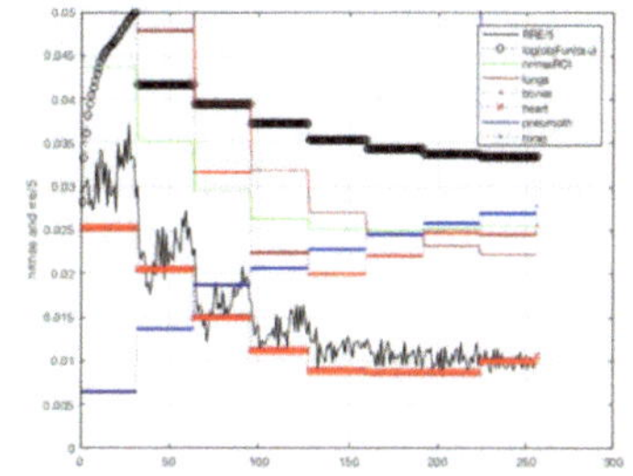

(f) Single transmission, TV, RMSE

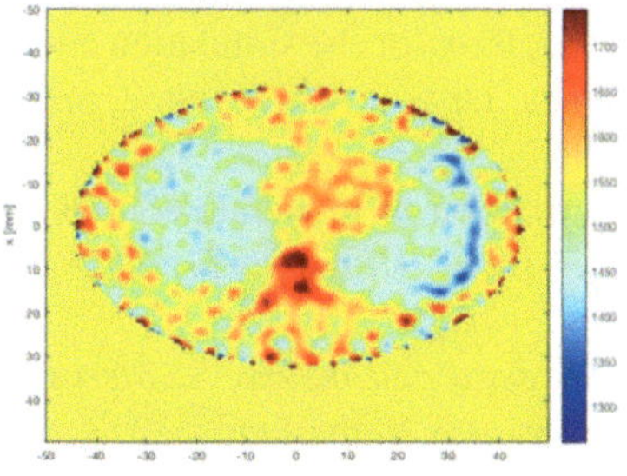

(g) Single transmission, Norm2Lapl $\alpha = 0.5$

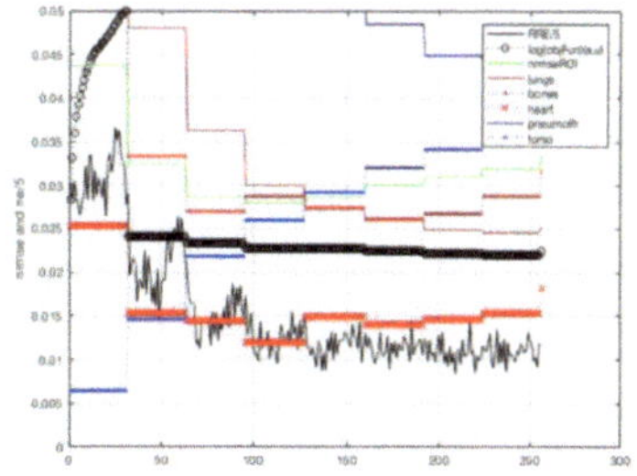

(h) Single transmission, Norm2Lapl, RMSE

Figure 8: Left column: Reconstructions after 8 iterations (full sweeps) of the phantom with a small pneumothorax in the left lung. Right column: RMSE by organ for each case.

Acknowledgements

The project was supported by Award Number R21EB009508 from the National Institute Of Biomedical Imaging And Bioengineering. The content is solely the responsibility of the authors and does not necessarily represent the official view of the National Institute Of Biomedical Imaging And Bioengineering or the National Institutes of Health. This study was also financed in part by the Coordenação de Aperfeiçoamento de Pessoal de Nível Superior - CAPES and FAPESP (Brazil).

References

[1] S. Mitragotri, "Healing sound: the use of ultrasound in drug delivery and other therapeutic applications," *Nature Reviews Drug Discovery*, vol. 4, pp. 255–260, 2005.

[2] F. Ahmadi, I. V. McLoughlin, S. Chauhan, and G. ter Haar, "Bio-effects and safety of low-intensity, low-frequency ultrasonic exposure," *Progress in Biophysics and Molecular Biology*, vol. 108, no. 3, pp. 119 – 138, 2012.

[3] I. V. McLoughlin, *Applied Speech and Audio Processing*. Cambridge Univ. Press, 2009.

[4] H. Dyball, "Talking ultrasound," *IIET Electronic Letters*, vol. 46, no. 6, p. 383, 2010.

[5] C. L. Tobin, Y. C. Lee, F. Gleeson, D. Feller-Kopman, and N. Rahman, *Pleural ultrasound for clinicians: a text and e-book.* CRC Press, 2014.

[6] R. D., H. H.-P., D. D., Z. P., and U. S., "Low-frequency ultrasound permeates the human thorax and lung: A novel approach to non-invasive monitoring," *Eur. J. Ultrasound*, vol. 31, p. 53, 2010.

[7] K. Morenz, H. Biller, F. Wolfram, S. Leonhadt, D. Rüter, T. Glaab, S. Uhlig, and J. M. Hohlfeld, "Detection of air trapping in chronic obstructive pulmonary disease by low frequency ultrasound," *BMC Pulmonary Medicine*, vol. 12, no. 8, 2012.

[8] B. E. Treeby and B. T. Cox, "k-Wave: MATLAB toolbox for the simulation and reconstruction of photoacoustic wave fields," *J. of Biomed. Optics*, vol. 15, no. 2, pp. 1 – 12, 2010.

[9] M. L. O. R. J. Lavarello, "A study on the reconstruction of moderate contrast targets using the distorted Born iterative method," *IEEE Trans. Ultrason. Ferroelectr. Freq. Control*, vol. 55, pp. 112–124, 2008.

[10] R. J. Lavarello and A. J. Hesford, *Methods for Forward and Inverse Scattering in Ultrasound Tomography*, pp. 345–394. Dordrecht: Springer Netherlands, 2013.

[11] F. A. Duck, "Chapter 4 - acoustic properties of tissue at ultrasonic frequencies," in *Physical Properties of Tissues* (F. A. Duck, ed.), pp. 73 – 135, London: Academic Press, 1990.

[12] T. D. Mast, A. Nachman, and R. C. Waag, "Focusing and imaging using eigenfunctions of the scattering operator," *J. Acoust. Soc. Am.*, vol. 102, no. 2, pp. 715–725, 1997.

[13] M. Cheney and D. Isaacson, "Distinguishability in impedance imaging," *IEEE Transactions on Biomedical Engineering*, vol. 39, no. 8, pp. 852–860, 1992.

[14] R. E. Appel, “Possibility of microcavitation from diagnostic ultrasound,” *IEEE Transactions on Ultrasonics, Ferroelectrics, and Frequency Control*, vol. 33, no. 2, pp. 139–142, 1986.
[15] B. M. U. S. (BMUS), *Guidelines for the safe use of diagnostic ultrasound equipment.* Tech. Rep., 2009.
[16] U. D. of Health, H. S. C. for Devices, and R. health, *Information for manufacturers seeking marketing clearance of diagnostic ultrasound systems and transducers, No. 510k.* Tech. Rep., 2008.
[17] F. Ahmad and I. V. McLoughlin, “A new mechanical index for gauging the human bioeffects of low frequency ultrasound,” 2013.

Life-like Phantoms for Biomedical Applications

A. Wydra[1] and R. Maev[1]

[1] *True Phantom Solutions Inc., E-mail: awydra@truephantom.com*

Abstract

In the various stages of developing diagnostic and therapeutic military and civilian equipment, the use of synthetic tissue mimicking materials can play a pivotal role in improving the process, help in implementation, testing and calibrations and in training paramedics, nurses and new doctors. The human-like structures made out of those synthetic tissues are called phantoms, which depending on their application (i.e. MRI, Ultrasound, or CT Scan), have to be made out of specific materials to match the respective physical characteristics of the tissues they have to mimic. In this paper we introduce imaging phantoms for ultrasound, MRI and CT. The authors show that by choosing the appropriate material, we can create phantoms which are able to mimic most of the desired properties of healthy and unhealthy biological tissues. The paper shows the examples of the fabricated phantoms, their properties and their comparison with the corresponding real tissues. In contrast to ex vivo tissues, the proposed phantoms can be custom designed and maintain their properties unchanged for long periods of time.

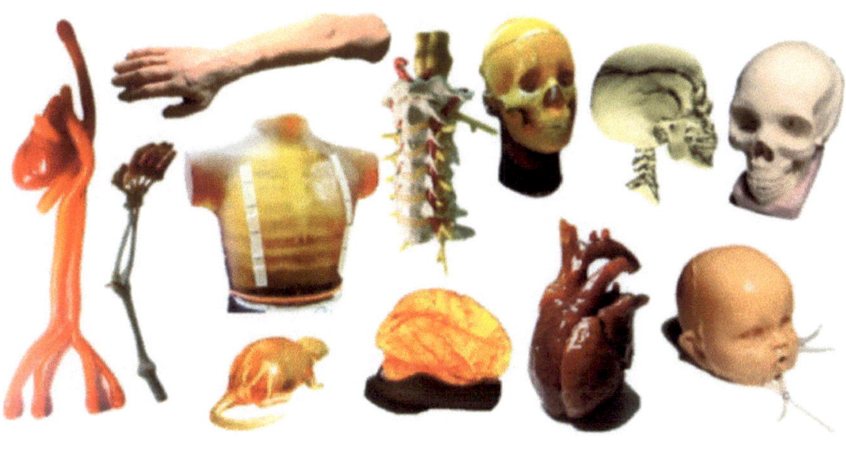

Quantitative Assessment of Skin using High-Resolution Handheld Ultrasonic Scanner

F. Seviaryn[1], G. Schreiner[1], S. Youseef[1], A. Maeva[2], I. Seviaryna[1], and R. Maev[1]

[1] *University of Windsor, Canada, E-mail: seviazy@uwindsor.ca*
[2] *University College London, United Kingdom*

Abstract

High-resolution ultrasonic imaging has been increasingly used in dermatology as a complementary technique for cutaneous lesions assessment. It may reduce the number of invasive procedures such as biopsies and assist in surgery planning. The success of ultrasonic methods highly relies on the ability of modern imaging systems to deliver precise and interpretable information. The new portable high-resolution ultrasonic imager was designed to be used for dermatological applications. The portable scanner operates at 50-100 MHz and provides B-scan images with up to 4 mm depth penetration with an axial resolution of 20 μm. Adaptive signal processing algorithms allow highlighting elements and features of the tissues. The primary skin layers, as well as the skin appendages, are identified on the obtained acoustical images. The benign melanocytic lesions (ephelides and nevi), skin burns and scars have been identified and characterized by their acoustic properties. Measurements of the thickness of skin layers at a different part of the body provide a base for statistical analysis and give information about skin conditions.

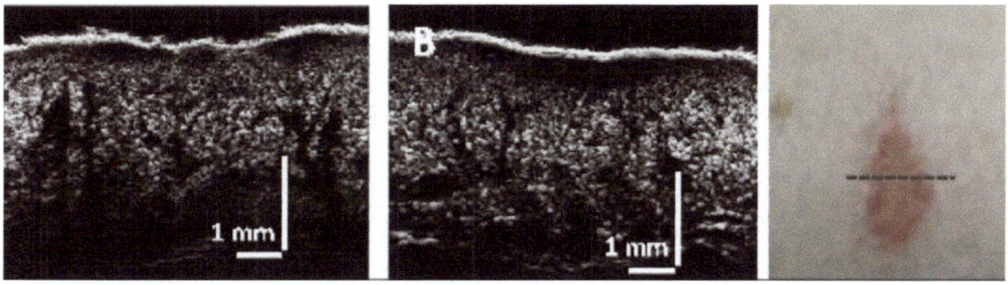

Figure 1: Ultrasonic images skin burn on the forearm. A. One day after the burn occur B. Seven days after. C. Orientation of the cross section across the lesion.

Characterization of skin morphology based on elastic modality can potentially estimate the overall condition of the skin and be used for diagnostic purposes. The new device is intended for the following dermatological applications: surgery planning and image-guided intervention, assessment of wound healing and skin grafts, various vascular anomalies, inflammatory diseases, and numerous cosmetic complications.

High-Resolution Mapping of Changes in Properties in Dermal Collagen Due to Light Damage

A. Maeva[1], I. Seviaryna[2], C. Hopper[1], C. Perrett[3], O. Levine[1], and A. Akbar[1]

[1] *University College London, United Kingdom, E-mail: anna.maeva.15@ucl.ac.uk*
[2] *University of Windsor, Canada*
[3] *University College London Hospital, United Kingdom*

Abstract

Age-related changes affect both the function and structure of human organs such as skin. Collagen plays a crucial role in providing elastic and mechanical strength to human skin. It is essential to understand and characterize any changes in collagen structure and properties related to ageing using a minimally invasive method. Mechanical properties of living tissues and cells like elasticity, viscosity and density directly associate with cell function.

One efficient method to evaluate tissue's mechanics is scanning acoustic microscopy (SAM), which allows non-invasive qualitative and quantitative assessment of tissue's stiffness without significant tissue preparation or damage. Porcine skin model was used to evaluate dermal collagen damage post-treatment with an Intense Pulse Light (IPL). Backneck fold biopsies from a 4-week-old, 25 kg pig were irradiated at increasing doses of 40 J/cm^2 once, thrice and ten times. Samples were then cryo-sectioned at a thickness of 10 μm. Ex-vivo biopsies were assessed with scanning acoustic microscope (Honda AMS-50SI equipped with 320 MHz focused transducer), both imaging atomic force microscopy (AFM) and mechanical AFM, and polarized light microscopy.

It was demonstrated that collagen fibres degrade depth-wise into the reticular dermis. At maximum irradiation, loss of banding and gelatinization was observed. Sound attenuation decreases with higher exposure. Damaged collagen demonstrated much lower sound speed and attenuation compared to healthy controls. Sound speed decreases at much faster rates than the attenuation. The ability of high-frequency ultrasound to quantitatively assess biological processes in the tissue places the method to a molecular level diagnostic device. Mapped sound speed and attenuation along with tissue and cells morphology significantly contribute to diagnostic imaging and can be a valid biomarker to assess alignment and state of collagen, which leads to ageing and inflammatory processes.

Poster papers and abstracts

Whole Breast Stiffness Imaging using Bulk Modulus by Ultrasound Tomography offers new Clinical Insights

P. J. Littrup[1-4], N. Duric[2,3], C. Li[3], M. Sak[3], and R. Brem[4]

[1] *Prof. of Radiology, Wayne State University, Detroit, MI, USA, E-mail: plittrup@delphinusmt.com*
[2] *Prof. of Oncology, Wayne State University, Detroit, MI, USA*
[3] *Delphinus Medical Technologies, Inc., Novi, MI, USA*
[4] *George Washington University School of Medicine & Health Sciences, D.C., USA*

Abstract

We described and tested the first clinically relevant, whole breast imaging of tissue stiffness using ultrasound tomography (UST) to map the bulk modulus differences in breast tissues, as well as benign and malignant masses. UST stiffness measurements extracted information on the tissue bulk modulus and were grouped by K-means clustering into 3 groups (i.e., soft, intermediate and stiff), while assessing impacts of spatial filtering upon imaging displays of focal mass stiffness. UST stiffness measurements demonstrated that 11.2% of total breast volume fell within the highest stiffness category, of which 80% (9.0/11.2) was confined to dense parenchyma. Yet, 68% of dense parenchyma was soft, such that only 9.0% total breast volume was stiff dense parenchyma. All smaller masses (<1.5 cm) showed significantly greater percentage of stiff components and were better represented by spatial filtering than their corresponding larger (>1.5 cm) masses ($p<0.001$). Cancers had significantly greater mean stiffness indices and mean homogeneity of stiffness ($p<0.05$). Whole breast stiffness imaging at sub-mm resolution suggests the feasibility of identifying small stiff masses, mainly due to cancers, amongst predominantly soft dense parenchyma. Further studies are needed to define the role of stiffness imaging for breast cancer treatment monito-ring, risk evaluation and screening women with dense breasts.

Keywords: ultrasound tomography, breast density, sound speed, breast cancer risk

1 Introduction

Palpation has been used for hundreds of years to detect pathologic changes in tissue stiffness, particularly for breast cancer (BC) diagnosis and screening. Yet, the well-established limitations of the clinical breast examination in sensitivity for BC detection (e.g., 54%) [1], as well as specificity in many women with normal variants, has limited its role. Palpation is more sensitive for suspicious stiff masses near the skin surface but can lead to false positives with stiff parenchyma or cysts, while deeper masses are frequently not palpable.

1.1 Problems with ultrasound elastography?

Mathematical concepts of physical stiffness encompass complex definitions/equations that simplify the variable parameters describing the deformation of a uniform material to load or stress [2], which doesn't accurately model the heterogeneity of complex breast tissue [3].

Handheld ultrasound (HHUS) evaluates elastic properties, commonly referred to as elastography, but only in a single dimension. HHUS thus only assesses localized tissue stiffness, or elasticity, while current automated breast US when use for screening does not incorporate elastography or measure stiffness. Yet, localized stiffness has been assessed by elastography to help characterize differences in benign and malignant masses [3-6], but not for BC screening. Therefore, tissue strain in axial and perpendicular planes approximate elastic (i.e., Young's) and shear moduli during HHUS elastography, respectively. But soft tissues do not have a simple mechanical nature which makes characterizing their elastic behavior with a single parameter. Since palpation produces multidimensional stress on target tissues, it may be more like a 3D strain parameter, or bulk modulus [3-6].

The bulk modulus expresses tissue properties as a material's resistance to uniform compression and associated volume changes. Bulk modulus values also have larger dynamic ranges than either Young's or shear moduli, which may facilitate improved tissue differentiation. For example, the metastatic potential of BC has shown strong inverse correlations with bulk tumor stiffness in animal models, which likely relate to tissue reactions of the surrounding extracellular matrix, producing less metastases and greater collagen in stiffer tumors [7]. Moreover, multi-parametric use of sound speed, attenuation and backscatter coefficient produced better separation of hepatic fibrosis in vitro but has not reached clinical application [8].

Women with dense breasts have greater cancer risk [9,10], which is compounded by detecting suspicious masses that are similarly dense on mammography. Stiffness has been shown to be an independent risk factor, separate from breast density, when it was volume-averaged over the whole breast by mammography [11]. Yet, stiffer components of dense parenchyma were not delineated, such that stiffness distributions throughout each breast were not mapped. Whole breast stiffness cannot be adequately performed by standard US using localized elastography and has only been reported on a limited basis for breast magnetic resonance (MR) [12,13]. Higher breast stiffness was noted by MR elastography in patients with denser breasts than those with lower breast densities, but was not assessed on a per patient basis [12,13].

1.2 Ultrasound elastography, density and stiffness

Ultrasound tomography (UST) has been developing over 40 years and recently accelerated with greater computer capacity [14-25]. Our research team helped develop a ring array, UST transducer for whole breast and focal mass evaluation, using quantitative transmission properties of sound speed (SS) and attenuation (ATT), combined with circumferential reflection [26-36]. Our UST experience demonstrated excellent correlation of SS with mammographic breast density [29-36], including better correlation with MR parenchymal distribution [37] and recent improvements in SS resolution [34]. Despite SS being a stronger BC risk factor than mammographic density [32], SS does not measure stiffness as an independent risk factor for BC [11].

In a limited study of 11 patients, we originally introduced and validated UST stiffness images in phantom studies and compared them to conventional elastography [39]. We have since developed *Whole Breast Stiffness Imaging* (WBSI) as a clinical tool for mapping focal stiffness throughout the breast, including the characterization of parenchyma, benign and malignant masses. Despite several other investigations of UST, the concept of stiffness was not considered, limiting their assessment to breast density [36], normal tissue classifications [40] and basic differentiation of cystic from solid masses [41,42].

WBSI by UST localizes gradients of stiffness throughout the breast and within masses [38,39]. WBSI may thus improve BC risk evaluation, conspicuity and diagnostic assessments of stiff regions within high SS dense parenchyma during screening. Before these applications can be thoroughly explored, both whole breast and mass evaluation by UST requires a basic understanding of the relative percentages and distributions of local stiffness, which also impacts the qualitative appearance to radiologists.

The optimal UST display of WBSI was initially assessed using both quantitative volumetric assessments of relative breast stiffness, as well as their qualitative appearances. Insight gained by these analyses may help guide UST applications for dense breast screening and/or mass characterization. If successful, future considerations include additional stiffness volume parameters for computer-aided detection, machine learning and/or diagnosis. Quantitative and qualitative aspects of both whole breast and mass stiffness imaging were separately considered as subsections for organizational clarity.

2 Methods

2.1 Patients and clinical evaluations

A total of 206 patients underwent informed consent for additional UST imaging as part of their routine clinical evaluation of a mammographic or palpable abnormality. This HIPAA protected, multicenter trial of UST (SoftVue, Delphinus Medical Technologies Inc., Novi, Michigan; Clinicaltrials.gov – NCT#02977247) containing at least 298 masses (Table 1). Whole breast or mass stiffness distributions included both quantitative and qualitative assessments.

Unless considered a characteristic cyst by standard HHUS evaluation, all masses underwent biopsy confirmation and separation by size (< and >1.5 cm diameter, Table 1). Breast density by mammography was noted as heterogeneously or extremely dense 64.6% (N = 133) and 26.7% (N = 55), respectively. In order to sample more cancers, 8.7% (N = 18) of women were included with scattered breast density.

Mass Type	#	< 1.5 cm	> 1.5 cm
Cancer	90	79	11
Fibroadenoma	106	88	18
Cyst	78	52	26
Other benign	24	21	3
Totals	**298**	**240**	**58**

Table 1: Mass type and size distributions ("Other benign" includes biopsy proven outcomes such as fibrosis, fibrocystic change, etc.)

2.2 UST and bulk modulus

Whole Breast Stiffness Imaging (WBSI) by UST included a surrogate of bulk modulus by UST and used multi-parametric comparisons of image stacks. No absolute physical unit values were generated for stiffness and quantitative values were displayed by a relative color scale.

UST Stiffness - Derivation: Longitudinal acoustic waves propagate according to

$$c \propto \sqrt{\frac{B}{\rho}},$$

whereby B represents bulk modulus (stiffness parameter) and ρ as density. The speed of sound, c and tissue density, ρ, are linearly correlated [48], such that:

$$c = 1.12\rho + 0.391,$$

a relationship noted from our previous studies of breast density [29-35]. Combining the above yields the proportionality:

$$B \propto c^3.$$

The bulk modulus of tissue thus has a one to one correspondence to the speed of sound. However, the dependence of SS upon B is weak:

$$c \propto B^{1/3}.$$

However, this relationship breaks down in water and only applies to solid tissues. Therefore, SS provides inaccurate stiffness estimates in certain breast pathologies, such as simple cysts.

In order to improve the robustness of stiffness, we introduced tissue attenuation as an additional measurable parameter. Attenuation also scales with density [48] but yields values close to zero for cysts (i.e., non-attenuating). Attenuation scales with tissue density as:

$$\alpha \propto 3.38\rho - 2.98.$$

Since the above has significantly lower correlation coefficient [48], we constructed a parameter, ζ, that is the product of attenuation and sound speed, αc. Combining all of the above relationships results in:

$$\zeta = \alpha c \propto \rho^2 \propto c^2 \propto B^{2/3}.$$

Then, expressing compressibility in terms of the measurables, we can reduce to:

$$B \propto \zeta^{3/2}.$$

Therefore, ζ combines the desirable properties of: (i) functional sensitivity to stiffness and (ii) yields a value of $B = 0$ when measured attenuation, α, is 0. Image processing then displays the bulk modulus, B, on a color scale. Color coding of the stiffness maps should not be interpreted in absolute units since no calibrations are being carried out. Bulk modulus values remain quantitative and the scale does not vary between patients.

2.3 Whole Breast Stiffness Distribution

Details have been described for UST quantification of dense and non-dense tissue components of whole breast volumes [29-37]. However, the distributions of stiffness sub-components were not assessed. This series includes different mass types, such that the overall volume stiffness distribution was considered with and without each mass' volume stiffness components. WBSI was reviewed for all patients and masses encountered in this clinical study. WBSI extracted information on the tissue bulk modulus and converted to an index of relative tissue stiffness in MIM for qualitative evaluation (MIM Software Inc, Cleveland, OH) [39]. WBSI provided a quantitative measure, such that the absolute value cannot be applied across all patients (i.e., no actual stiffness units), which may relative color mapping of bulk modulus within each patient's breast more appropriate for volume distributions of stiffness subcategories. K-means clustering techniques, similar to techniques used for generating volume averaged sound speed [31-33,35,36], determined the thresholds for 3 levels of relative stiffness (i.e., stiff, intermediate and soft), and then applied to whole breast image stacks. The 3 levels of stiffness generated average volumes for the stiffest, intermediate and soft tissues, relative to the volume averaged SS. Stiffness distributions were calculated for each patient and then reported as an average percentage for all patients. Total breast volume for each patient first included underlying associated masses that were then subtracted from the total volume, as well as their mass stiffness sub-volumes. The relative incidence of each breast mass category was then produced for the six different, non-mass breast volumes of: stiff, intermediate and soft distributions within both dense and non-dense total breast tissue.

2.4 Mass Stiffness Distribution

Mass margins were hand-traced by a radiologist with over 20 years of UST experience and breast imager (PL) using MIM software to generate regions of interest (ROI) surrounding all masses detected by UST. The single best visualized/representative image upon a combination of sound speed and reflection image stacks were used to traced mass margins, generating a surface area. Stiffness indices of mass ROIs were compared with the proportions of density (SS) and stiffness, determined by K means clustering. Percentages of these stiffness components in relation to dense and non-dense mass components were also graphically compared. Similar to whole-breast volume evaluations, *qualitative observations* of mass stiffness provided clinical context to the quantitative volume data, thereby entering the realm of texture evaluation of stiffness within masses. To compare the potential of quantitative texture evaluations over absolute stiffness values, the mean homogeneity of stiffness was calculated for each mass using the Gray-Level Co-Occurrence Matrix (GLCM) approach [49,50]. The ability to differentiate masses using stiffness parameters was assessed statistically using the t-test. Significance was based on a p value < 0.05.

3 Results

3.1 Quantitative - Whole Breast Stiffness

Considering the high incidence of cysts in women with dense breasts, there were 1.4 masses per woman (298/206) or 1.2 (239/206) masses per SoftVue scanned breast. The relative volume of dense tissue by SS versus stiff tissue showed a correlation coefficient of 0.0048, suggesting that the volumetric stiffness distribution was independent of tissue density.

	Dense	Non-dense	Total
Stiff	9.0%	2.2%	**11.2%**
Intermediate	11.8%	21.7%	**33.4%**
Soft	8.3%	47.1%	**55.4%**
Total	**29.1%**	**71.0%**	**100.%**

Table 2: Whole breast volume distributions of stiffness components for dense and non-dense (fatty) tissues, including their underlying masses (N = 239 breasts).

The distribution of the three stiffness categories according to dense and non-dense, or fatty, tissues are noted in Table 2. Masses were not excluded from the total breast volume due to minimal volume contribution to stiffness distributions (i.e., average breast vol. = 737cc, compared to average tumor volume = 1.1cc). Only 32% (9.0/29.1) of dense parenchyma had high stiffness (i.e., red), while 80% (9.0/11.2) of this high stiffness resided within dense parenchyma (Table 2). Conversely, most dense parenchyma (68%) was soft, while 97% (68.8/71) of non-dense (fatty) tissue was soft. The 2.2% of stiff fat could be artifactual since K-means clustering may overestimate the non-fat pixels adjacent to dense parenchyma.

Qualitative - WBSI: Figure 1 displays SS and unfiltered Stiffness Fusion images of the three breast densities groups encountered in this study. Dense tissue is represented by white components in SS images, while corresponding stiffness data is shown in the form of a color map overlaid on a grey-scale reflection image, providing underlying tissue anatomy. A comparison of the SS and stiffness images shows that some dense tissues can be stiff, soft or a mixture of both.

Figure 2 shows SS and both filtered/unfiltered images of a mammographically occult cancer. The unfiltered stiffness visualization (Figure 2D) is partially obscuring the cancer by colorization of the adjacent dense and stiff parenchyma. The cancers mass effect is actually better defined by SS and the spatially filtered stiffness rendering (Figure 2E)(see also in Masses to follow). In this example, there is a cancer at 5-6 o'clock sitting on the edge of stiff dense tissue. While the unfiltered stiffness image (Figure 2D) characterizes the overall stiff tissue distribution in the breast, the filtered stiffness image (Figure 2E) more strongly highlights the suspicious mass by suppressing the red contributions (stiff regions) from the normal stiff breast parenchyma and skin. The cancer lies along the fat-glandular interface, a characteristic that has been noted for 94-99% cancers by breast MRI [38, 43-45], and especially notable when visualized in the native coronal plane of UST.

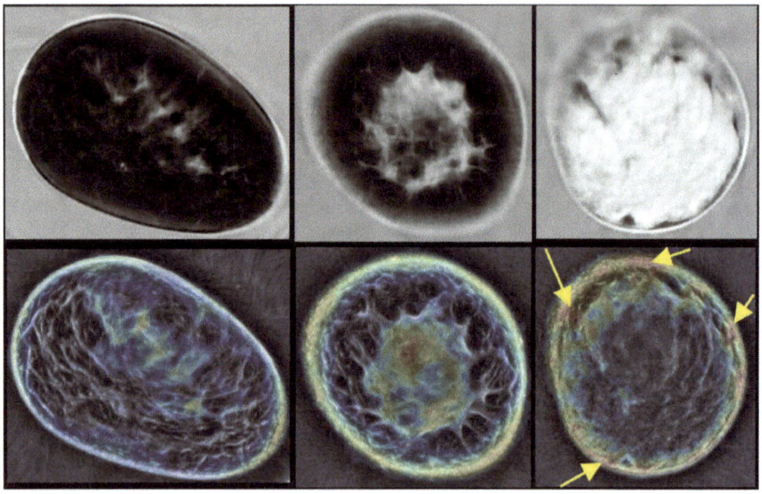

Figure 1: Most dense tissues are soft,while the stiffest tissues make up a small percentage of total breast volume. Top row: SS images show denser tissues (white) from scattered (left), to heterogeneously (middle), and extremely dense (right). Bottom row: Corresponding unfiltered stiffness images show their distribution, with blue/black representing the softest tissues and stiffness increasing from green/yellow (intermediate) to red (stiffest). Skin can show artifactual stiffness due to attenuation component and/or refraction (arrows).

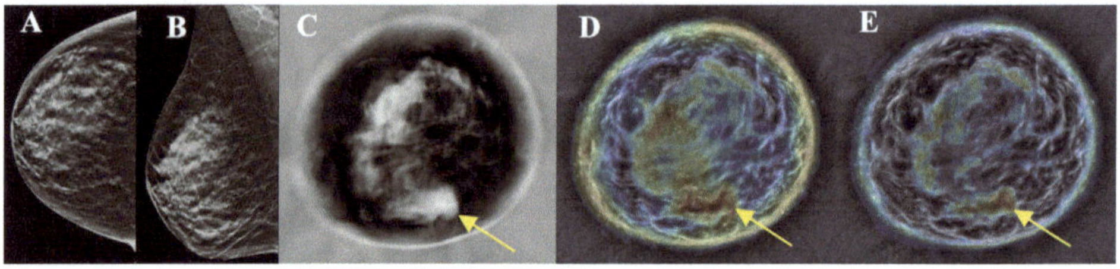

Figure 2: Cancer patient: Mammograms from a woman with heterogeneously dense breasts in cranio-caudal (A) and medio-lateral oblique (B) projections are compared to UST SS image (C), unfiltered (D) and filtered stiffness (E) images. The unfiltered stiffness image (D) shows a large red area at 6:00 that partially obscures the underlying mass effect from the cancer (arrows) at 5:00, better seen on SS (C) and filtered stiffness (E) images.

3.2 Quantitative - Mass Stiffness Distributions

Figure 3 shows examples of the filtered WBSI versions, showing small and large cysts, fibroadenomas and cancers. Cysts generally had soft appearances (blue) with little or no internal stiffness, regardless of size (Figure 3A,B). Fibroadenomas can have either homogenous or mildly heterogenous appearance and can be either soft, stiff or mixed (Figure 3C,D). Cancers, on the other hand, tend to be largely heterogenous, and very stiff (red color), especially when small (Figure 3E,F).

Optimal visualization of stiffness for masses was explored for the unfiltered and filtered stiffness algorithms. Smaller masses (i.e., $<$1.5 cm) had significantly greater percentages of the stiffest component (i.e., red), regardless of filtering algorithm or tumor type. Alternatively, larger masses (i.e., $>$1.5 cm) had significantly greater percentages of the softer components (chi-squared; $p < 0.001$).

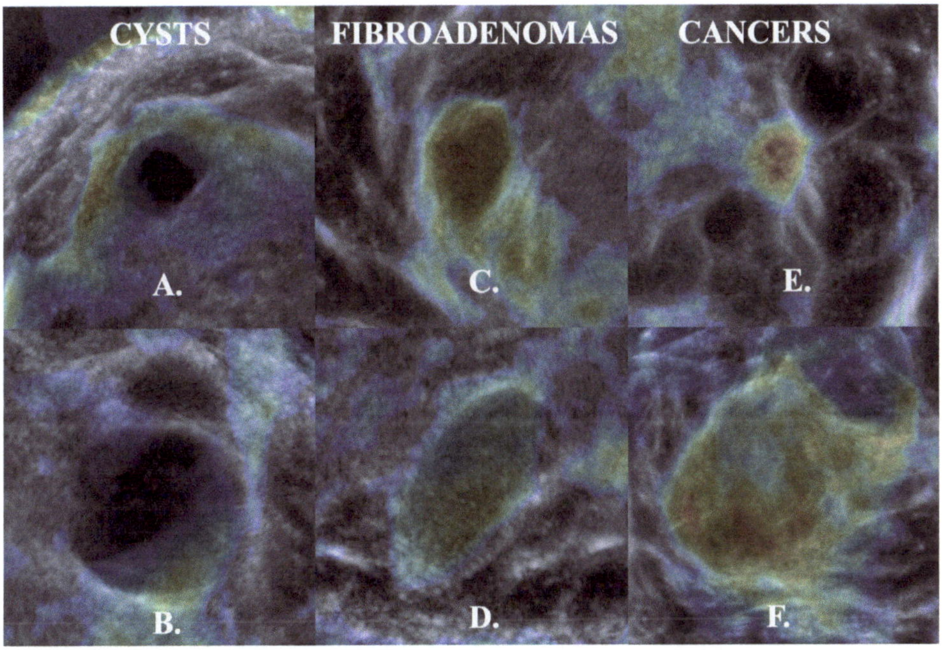

Figure 3: Cysts (A,B), Fibroadenomas (C,D), and Cancers (E,F). Top row shows masses $<$1.5 cm. Bottom row shows comparable masses $>$1.5 cm.

For the filtered algorithm, small cancers reached greater statistical significance for percentage stiff component, compared with fibroadenomas (t test, $p < 0.001$). Larger masses had lower percentage stiff components, which would be expected with spatial filtering.

Table 3 compares the stiffness values of large and small masses, along with the corresponding statistical test results. The smaller ranges of stiffness and less overlap for the filtered rendering produced significantly greater discrimination of smaller cancers from fibroadenomas (i.e., $p = 0.00036$ versus $p = 0.08$). Conversely, the unfiltered stiffness images better separated the larger cancers from fibroadenomas ($p = 0.037$ versus $p = 0.127$).

Finally, stiffness indices and homogeneity values included all masses and showed similar distributions for both filtered and unfiltered stiffness images. Statistically, stiffness index and homogeneity value differences between the mass types are significant for both filtered and unfiltered stiffness images, respectively ($p<0.05$).

Table 3: Tumor Comparisons			**Stiffness (5%-95% C.I)**		**p values**
Small	**Unfiltered**	Cancer (CA)	0.1256-0.8585	CA vs Cyst:	1E-09
≤ 1.5 cm		Fibroadenoma (FA)	0.0021-0.8634	**CA vs FA:**	**0.08**
		Cyst	0.00012-0.6499	Cyst vs FA:	0.0000017
	Filtered	*Cancer (CA)*	*0.1247-0.6967*	*CA vs Cyst:*	*6.00E-11*
		Fibroadenoma (FA)	*0.1704-0.5303*	***CA vs FA:***	***0.000036***
		Cyst	*0.001-0.3618*	*Cyst vs FA:*	*0.00028*
Large	**Unfiltered**	Cancer (CA)	0.1916-0.8288	CA vs Cyst:	2.00E-10
> 1.5 cm		Fibroadenoma (FA)	0.036-0.7005	**CA vs FA:**	**0.037**
		Cyst	0.00034-0.2364	Cyst vs FA:	0.00013
	Filtered	*Cancer (CA)*	*0.0826-0.8104*	*CA vs Cyst:*	*0.00000021*
		Fibroadenoma (FA)	*0.0197-0.4830*	***CA vs FA:***	***0.127***
		Cyst	*0.00005-0.0691*	*Cyst vs FA:*	*0.000025*

Table 3: Mass comparisons according to size and stiffness values using unfiltered and filtered stiffness values. The filtered algorithm accentuates smaller masses, producing significantly better differentiation of cancers from fibroadenomas, whereas that spatial filtering also degrades performance for those larger solid masses.

3.3 Qualitative - Mass Stiffness Imaging

Histologically, some smaller cysts containing stiffer components were commonly associated with complex cyst contents (i.e., by standard handheld US) and underwent aspiration/biopsy, whereas larger cysts were nearly all simple. Stiffness within fibroadenomas was generally associated with heterogeneous blending of the stiff component along the mass periphery.

Cancers had the greatest percentages of the stiffest component, whereby smaller cancers are predominantly stiff compared with larger cancers (Figure 3). Smaller cancers most often had their stiffest component centrally located (Figure 3E), whereas larger cancers often had an asymmetric clustered stiff portion (Figure 3F), rather than the heterogeneous blending noted for fibroadenomas (Figure 3C,D). The 4th histologic category of "other benign" had lower numbers (i.e., N = 24), with only 3 larger masses showing a stiffness pattern like cysts (i.e., predominantly soft), of which 2 showed fibrocystic change and 1 granulomatous mastitis. Most of the smaller, "other benign" category (i.e., N = 21) suggested similar stiffness distributions as cancer with histology mostly showing underlying fibrosis (i.e., biopsy report descriptions).

4 Discussion

Our WBSI data represents the first quantification and mapping of relative stiffness percentages of dense and non-dense tissue (i.e., fibroglandular/stroma and fat, respectively) from chest wall to nipple, while offering clinical insights for available masses. We acknowledge that these are early WBSI evaluations and present only initial descriptions and interpretations for both whole breast and mass evaluations. Few papers have addressed relative percentages of tissue stiffness, other than direct measurements of resected specimens [44]. Their paper did not address cysts but showed a progressive increase in stiffer components from benign tissues to different types of cancer. Qualitative aspects of WBSI described imaging appearances of stiffness encountered throughout the whole breast, as well as stiffness patterns within masses. These are prone to human interpretation bias, but augmented the quantitative analyses, in order to facilitate current physician training, interpretation and/or future analyses by machine learning. The *quantitative* subheadings for whole breast and mass evaluations are considered first.

4.1 Quantitative WBSI

As visualized in Figure 2, dense white tissue by SS can be decomposed into stiff, soft and mixed stiffness components. These images were able to quantitatively separate dense breast parenchyma from fat by grouping both into stiff, intermediate and soft categories by K-means clustering (see Materials and Methods section). While 80% of the stiffest category occurred in dense parenchyma, only 32% of dense parenchyma was actually stiff. Conversely, 68% of dense parenchyma was relatively soft, which potentially limits the presumed higher risk stiff-dense tissue to 9.0% of the breast volume (Table 2).

We demonstrated volumetric decoupling of stiffness from density, which has implications for BC risk evaluations, as well as dense breast screening. WBSI offers greater accuracy for both, by stratifying and localizing stiff tissues throughout the breast, compared with the more global mammographic compression technique [11]. We have shown that UST sound speed is a risk factor for developing BC and is more strongly associated with BC risk than mammographic percent density. The independence of stiffness from density could further provide additional risk criteria in support of clinical risk models. Our study demonstrates that stiffness can be measured throughout the uncompressed breast at high spatial resolution (< 1 mm). More accurate reduction of risk at the individual level would spur research into new chemoprevention treatments, as well as provide guidance for use of existing treatments such as tamoxifen and/or dietary changes.

4.2 Quantitative Mass Evaluation

Application of WBSI leads to the differentiation of cancer stiffness characteristics from benign masses (Figure 3). For masses < 1.5 cm, this differentiation significantly improved with spatial filtering (table 3). Masses <1.5 cm had significantly greater percentages of the stiffest component than larger masses masses [i.e.,>1.5 cm ($p<0.001$)]. The spatially filtered stiffness algoritm (i.e., structures >1.5 cm) increased the percentage of the stiffest component for smaller cancers by removing any coincident or confounding, larger stiff fibroglandular tissue. This enhancement likely leads to the unmasking of the true underlying stiffness of smaller tumors (Figures 2E and 3E), but can confound differentiation of larger solid masses (Table 3). The high stiffness of small cancers suggests a future role for whole breast filtered stiffness in BC screening, particularly in an adjunctive setting where cancers missed by mammography tend to be small invasive cancers, usually 0.5-1.5 cm. However, this paper addressed only known masses within a clinical series, whereas the potential for spatial filtering producing false positives will require further analyses from future screening cohorts.

4.3 Qualitative Evaluations

Table 3 shows that the improved mass differentiation by absolute stiffness values predominantly applied only to smaller masses. The ability to differentiate groups of cysts, fibroadenomas and cancers was highly significant. However, the overlap of their stiffness distributions limits the accuracy of the differentiation on a case by case basis (Tables 3). These results suggest that stiffness could be a valuable parameter for mass differentiation when combined with other criteria (e.g. BIRADS) in future machine learning scenarios.

Mass differentiation is also possible using texture features, such as the homogeneity of stiffness, which are more amenable to computer measurements than the human eye. Therefore, once a mass is identified, future computer-assisted detection/diagnostic algorithms, or machine learning with radiomics may better utilize a spatial filter to improve mass detection and/or characterization. In addition, these quantitative mass evaluations may have immediate application in monitoring cancer treatment response to neo-adjuvant chemotherapy over time.

Qualitative WBSI evaluations of the whole breast and mass stiffness renderings are prone to operator-dependent subjective impressions but offer insights to the potential opposing forces driving sensitivity and specificity. Our study does not represent a screening series whereby sensitivity is a major concern for mass detection. The ability to rapidly interpret WBSI still requires optimal detection of an underlying mass before characterization. Further work is needed with screening data to accurately balance the goals of sensitivity and specificity by different stiffness rendering regimes.

Qualitative mass stiffness assessments are subjective, but may have more immediate clinical implications than whole breast stiffness evaluations. Small cancers have been noted to be generally stiff, (Figure 2E, 3E), especially for ER/PR positive cancers with associated scirrhous reactions of the surrounding peritumoral region [38]. Larger cancers on the other hand may be prone to central necrosis (Figure 3F) or may have an undifferentiated tumor type (e.g., triple negative), or less peritumoral scirrhous reaction, both of which may confer lower bulk modulus, or softness, consistent with our finding that larger cancers are softer than smaller cancers. Both stiffness and homogeneity values were impacted by the spatially filtered algorithm (Tables 3), producing more focal clustering of redness (i.e., greater number of red pixels grouped together) and enabled texture considerations. Immediate clinical application of WGSI is thus possible for frequent UST monitoring of tumor stiffness response to neoadjuvant chemotherapy over time compared to breast MR with lower cost, no gadolinium injection, and shorter exam times.

Stiffness rendering of cancer also has pattern recognition implications beyond the initial percentage stiffness distributions. Larger cancers were softer, but showed clustering of red pixels along tumor margins, compatible with known peritumoral stiffness along an invasive margin arising from surrounding scirrhous reactions to local extraductal extension [38]. Similarly, fibroadenomas often had blended stiffness (Figure 3C/D), relating to their relative scirrhous component of whorled hyalinized regions blending with adenomatous tissue inside an encapsulated mass [46]. In contrast, cancers have greater internal "chaos", or heterogeneity (Figure 3E/F) than the low stiffness values and greater homogeneity of cysts (Figure 3A,B). Further work is needed on a UST imaging lexicon/library, which also has implications for future machine learning.

Weaknesses are inherent when using qualitative and quantitative stiffness descriptions of the new clinical breast imaging modality of UST. Limited insights can be made of potential UST application to dense breast screening. While the unfiltered stiffness algorithm appears well suited to whole breast evaluation in support of future BC screening and/or risk assessment, the filtered version may have more targeted application for improved mass detection and/or evaluation, but size dependent. We acknowledge that it is very early to presume optimal renderings for either screening or diagnostic uses of UST-WBSI.

UST technology is mainly limited by computing power such that continued improvements are expected for image rendering, computer-aided detection, machine learning and the multiple innate quantitative properties that can be analyzed from UST imaging. Whole breast, or focal mass, stiffness by UST may become a valuable parameter in support of cost-effective risk evaluations, BC detection, and/or neoadjuvant chemotherapy treatment monitoring.

5 Conclusion

Quantifying and mapping tissue stiffness throughout the whole breast is not currently available. This study also demonstrates that lesion characterization by WBSI is feasible. Using WBSI in a screening environment has the potential to reduce call backs and biopsies, in addition to other characteristics to improve specificity.

References

[1] M.B. Barton, R. Harris, S.W. Fletcher. The Rational Clinical Examination. Does This Patient Have Breast Cancer? The Screening Clinical Breast Examination: Should It Be Done? How? JAMA 282, 1270-1280 (1999).

[2] F. Baumgart. Stiffness - an unknown world of mechanical science? Injury 2000 May;31 Suppl 2:S-B14-23. Elsevier; 31: 14–84.

[3] P.N.T. Wells and H.-D. Liang. Medical ultrasound: imaging of soft tissue strain and elasticity. J. R. Soc Interface 8, 1521–1549 (2011) doi:10.1098/rsif.2011.0054

[4] J.F. Greenleaf, M. Fatemi, M. Insana. Selected Methods for Imaging Elastic Properties Of Biological Tissues. Annu Rev Biomed Eng 5:57–78 (2003).

[5] F.A. Duck. Physical Properties of Tissue. London, England: Academic Press; 1990.

[6] T. Glozman, H. Azhari. A Method for Characterization of Tissue Elastic Properties Combining Ultrasonic Computed Tomography with Elastography. Ultrasound Med 29, 387-98 (2010).

[7] J. Fenner, A.C. Stacer, F. Winterroth, T.D. Johnson, K.E. Luker, G.D. Luker. Macroscopic Stiffness of Breast Tumors Predicts Metastasis. Sci Rep. 4:5512. doi: 10.1038/srep05512 (2014).

[8] M. Meziri, W.C. Pereira, A. Abdelwahab, C. Degott, P. Laugier . In Vitro Chronic Hepatic Disease Characterization with a Multiparametric Ultrasonic Approach. Ultrasonics 43, 305-13 (2005).

[9] N. Boyd, L. Martin, S. Chavez, A. Gunasekara, A. Salleh, O. Melnichouk, M. Yaffe, C. Friedenreich, S. Minkin, M. Bronskill. Breast-Tissue Composition and Other Risk Factors for Breast Cancer in Young Women: A Cross-Sectional Study. The Lancet Oncology 10(6):569-580 (2009). Doi:10.1016/S1470-2045(09)70078-6

[10] N.F. Boyd, L.J. Martin, M. Bronskill, M.J. Yaffe, N. Duric, and S. Minkin. Breast Tissue Composition and Susceptibility to Breast Cancer. J Natl Cancer Inst 102(16): 1224-37 (2009).

[11] N.F. Boyd, Q. Li, O. Melnichouk, E. Huszti, L.J. Martin, A. Gunasekara, G. Mawdsley, M.J. Yaffe, S. Minkin. Evidence That Breast Tissue Stiffness Is Associated with Risk Of Breast Cancer. PLoS One. Jul 10, 9(7):e100937 (2014).

[12] J.R. Hawley, P. Kalra, X. Mo, B. Raterman, L.D. Yee, A. Kolipaka. Quantification of Breast Stiffness Using MR Elastography at 3 Tesla with A Soft Sternal Driver: A Reproducibility Study. J Magn Reson Imaging 45:1379-1384 (2017).

[13] A.E. Bohte, J.L. Nelissen, J.H. Runge, O. Holub, S.A. Lambert, L. de Graaf L, et al. Breast Magnetic Resonance Elastography: A Review of Clinical Work and Future Perspectives. NMR Biomed. 31:e3932. doi: 10.1002/nbm.3932, (2018).
[14] S. Johnson, D.T. Borup, J.W. Wiskin, M. Berggren, B. Hanover, F. Setinsek, et al. ., From Laboratory To Clinical Trials: An Odyssey Of Ultrasound Inverse Scattering Imaging For Breast Cancer Diagnosis. The Journal of the Acoustical Society of America. 120: p. 3023 (2006).
[15] S.A. Johnson and M.L. Tracy ML. Inverse Scattering Solutions By a Sinc Basis, Multiple Source, Moment Method. Part I: Theory, Ultrasonic Imaging, 5:361-375, (1983).
[16] J.S. Schreiman, J.J. Gisvold, J.F. Greenleaf, R.C. Bahn. Ultrasound Transmission Computed Tomography of The Breast. Radiology; 150:523-30, (1984).
[17] G.H. Glover. Computerized Time-Of-Flight Ultrasonic Tomography For Breast Examination. Ultrasound In Medicine & Biology. 3(2-3):117-27, (1977).
[18] F.A. Natterer. Propagation Backpropagation Method for Ultrasound Tomography, Inverse problems; 11:1225-1232, (1995).
[19] P.L. Carson, C.R. Meyer, A.L. Scherzinger, T.V. Oughton. Breast Imaging In Coronal Planes With Simultaneous Pulse Echo And Transmission Ultrasound. Science. 214(4525):1141-3, (1981).
[20] M. P. Andre, H.S. Janee, P. J. Martin, G. P. Otto, B. A. Spivey, D. A. Palmer, High-Speed Data Acquisition In A Diffraction Tomography System Employing Large-Scale Toroidal Arrays, International Journal of Imaging Systems and Technology. 8:137- 147, (1997).
[21] S.A. Johnson, D. T. Borup, J.W. Wiskin, F. Natterer, F. Wuebbling, Y. Zhang, C. Olsen Apparatus and Method for Imaging with Wavefields using Inverse Scattering Techniques. United States Patent 6,005,916 (1999).
[22] V.Z Marmarelis, T. Kim, R.E. Shehada, Proceedings of the SPIE: Medical Imaging 2003; San Diego, California; Feb. 23-28, 2002. Ultrasonic Imaging and Signal Processing – Paper 5035-6.
[23] D.L. Liu and R. C. Waag, Propagation and backpropagation for ultrasonic wavefront design, IEEE Trans. on Ultras. Ferro. and Freq. Contr. 44(1):1-13 (1997).
[24] H. Gemmeke and N. Ruiter. 3D Ultrasound Computer Tomography for Medical Imaging. Nuclear instruments and methods in Physics Research Section A: Accelerators, Spectrometers, Detectors and Associated Equipment, 580: 1057-1065, (2007).
[25] N. Ruiter, G. Göbel, L. Berger, M. Zapf and H. Gemmeke Realization of an Optimized 3D USCT, Proc. SPIE 7968, 796805 (2011)
[26] N. Duric, P.J. Littrup, L. Poulo, A. Babkin, R. Pevzner, E. Holsapple, O. Rama, and C. Glide. Detection of Breast Cancer With Ultrasound Tomography: First Results with the Computed Ultrasound Risk Evaluation (CURE) Prototype. Med Phys. 34: 773-85, (2007).
[27] C. Li, G.S. Sandhu, M. Boone, N. Duric, P. Littrup, M. Sak, and K. Bergman. Breast Tissue Characterization with Sound Speed and Tissue Stiffness. In Proceedings of the International Workshop on Medical Ultrasound Tomography: 1.-3. Nov. 2017, Speyer, Germany (p. 217). KIT Scientific Publishing (2018).

[28] B. Ranger, P.J. Littrup, N. Duric, P. Chandiwala-Mody, C. Li, S. Schmidt, J. Lupinacci. Breast Ultrasound Tomography versus MRI For Clinical Display of Anatomy and Tumor Rendering: Preliminary Results. AJR Am J Roentgenol; 198:233-9, (2012).

[29] C. Glide, N. Duric, P. Littrup: Novel Approach to Evaluating Breast Density Utilizing Ultrasound Tomography. Med Phys, 34:744-753, (2007).

[30] C.K. Glide-Hurst, N. Duric, P. Littrup: Volumetric breast density evaluation from ultrasound tomography images. Med Phys , 35:3988-3997, (2008).

[31] N. Duric, N. Boyd, P. Littrup, M. Sak, L. Myc, C. Li, E. West, S. Minkin, L. Martin, M. Yaffe, S. Schmid, M. Faiz, J. Shen, O. Melnichouk, Q. Li, and T. Albrecht. Breast Density Measurements with Ultrasound Tomography: A Comparison with Film and Digital Mammography. Med Phys; 40(1): 013501, (2013).

[32] N. Duric, M. Sak, S. Fan, R.M. Pfeiffer, P.J. Littrup, M.S. Simon, D.H. Gorski, H. Ali, K.S. Purrington, R.F. Brem, M.E. Sherman, G. Gierach. Using Whole Breast Ultrasound Tomography to Improve Breast Cancer Risk Assessment: A Novel Risk Factor Based on the Quantitative Tissue Property of Sound Speed. Journal of Clinical Medicine. 9(2): E367. doi: 10.3390/jcm9020367, (2020).

[33] Z.G. Khodr, M.A Sak, R.M. Pfeiffer, N. Duric, P. Littrup, L. Bey-Knight, H. Ali, P. Vallieres, M.E. Sherman, and G.L. Gierach. Determinants of The Reliability of Ultrasound Tomography Sound Speed Estimates as A Surrogate for Volumetric Breast Density. Medical Physics; 42(10), pp.5671-5678, (2015).

[34] G.Y. Sandhu, C. Li, O. Roy, S. Schmidt and N. Duric. Frequency Domain Ultrasound Waveform Tomography: Breast Imaging Using A Ring Transducer. Physics in Medicine & Biology; 60(14), p.5381 (2015).

[35] M. Sak, N. Duric, P. Littrup, L. Bey-Knight, H. Ali, P. Vallieres, M.E. Sherman, and G.L. Gierach. Using Speed of Sound Imaging to Characterize Breast Density. Ultrasound Med Biol, 43(1): 91-103 (2017).

[36] J. Wiskin, B. Malik, R. Natesan, M. Lenox. Quantitative Assessment of Breast Density Using Transmission Ultrasound Tomography. Med Phys. 46(6):2610-20, (2019).

[37] E.A.M. O'Flynn, J. Fromageau, A.E. Ledger, A. Messa, A. D'Aquino, M.J. Schoemaker, M. Schmidt, N. Duric, A.J. Swerdlow, and J.C. Bamber. Ultrasound Tomography Evaluation of Breast Density: A Comparison with Non-Contrast Magnetic Resonance Imaging. Invest Radiol 52(6): 343-348 (2017).

[38] P.J. Littrup, N. Duric, C. Li M. Sak, G. Sandhu, K. Bergman, et al. Current challenges in breast screening and diagnosis: From molecules to peritumoral regions and radiomics - The emerging imaging of whole breast stiffness. Proceedings of the International Workshop on Medical Ultrasound Tomography: 1.-3., Speyer, Germany (2017).

[39] N. Duric, P. Littrup, C. Li, O. Roy, S. Schmidt, J. Seamans, A. Wallen, and L. Bey-Knight, L., 2015, March. Whole breast tissue characterization with ultrasound tomography. In Medical Imaging: Ultrasonic Imaging and Tomography, 9419: 94190G (2015). International Society for Optics and Photonics.

[40] B. Malik, J. Klock, J. Wiskin and M. Lenox. Objective breast tissue image classification using quantitative transmission ultrasound tomography. Sci Rep; 6:38857 (2016).

[41] E. Iuanow, K. Smith, N. Obuchowski, J. Bullen and J. Klock. Accuracy of Cyst Versus Solid Diagnosis in the Breast Using Quantitative Transmission (QT) Ultrasound. Acad Radiol; 24(9):1148-1153 (2017).

[42] B. Malik and J. Klock. Breast Cyst Fluid Analysis Correlations with Speed of Sound Using Transmission Ultrasound. Acad Radiol; 26(1):76-85 (2019).

[43] J. Yamaguchi, H. Ohtani H, Nakamura K, Shimokawa I and Kanematsu T. Prognostic Impact of Marginal Adipose Tissue Invasion in Ductal Carcinoma Of The Breast. Am J Clin Pathol, 130(3):382-8 (2008).

[44] W.H. Kim, M. Li, W. Han, H.S. Ryu and W.K. Moon. The Spatial Relationship of Malignant and Benign Breast Lesions with Respect to the Fat-Gland Interface on Magnetic Resonance Imaging. Nature Sci Rep., 6:39085 (2016).

[45] W. Zhu, S. Harvey, K.J. Macura, D.M. Euhus, D. Artemov. Invasive Breast Cancer Preferably and Predominantly Occurs at the Interface Between Fibroglandular and Adipose Tissue. Clin Breast Cancer; 17(1):e11-e18 (2017).

[46] P.J. Littrup, L. Freeman-Gibb, A. Andea, M. White, K.C. Amerikia, D. Bouwman, T. Harb, W. Sakr. Cryotherapy for Breast Fibroadenomas. Radiology; 234: 63-72 (2005).

[47] E.V. Gubarkova, A.A. Sovetsky, V.Y. Zaitsev, A.L. Matveyev, D.A. Vorontsov, M.A. Sirotkina, et al. OCT-Elastography-based Optical Biopsy for Breast Cancer Delineation and Express Assessment of Morphological/Molecular Subtypes. Biomed Opt Express.; 10:2244-2263 (2019).

[48] TD Mast: Empirical Relationships Between Acoustic Parameters In Human Soft Tissues. ARLO, 1:37-42, (2000).

[49] K.S. Haralick and I. Dinstein, Texture Features for Image Classification, IEEE Transaction, Syst., Man., Cybern., vol SMC-3, 6 610-621, (1973).

[50] R.D. Chitalia and Kontos D. Role of Texture Analysis in Breast MRI as a Cancer Biomarker: A Review. J. Magn. Reson. Imaging; 49: 927-938.

Whole Breast Sound Speed Measurement from Ultrasound Tomography Correlates Strongly with Volumetric Breast Density from Mammography

M. Sak[1], N. Duric[1,2], P. Littrup[1], and R. Brem[3]

[1] *Delphinus Medical Technologies, Inc, E-mail: msak@delphinusmt.com*
[2] *Karmanos Cancer Institute*
[3] *George Washington University School of Medicine & Health Sciences, D.C., USA*

Abstract

Purpose

Quantitative sound speed volume measurements from ultrasound tomography (UST) have been shown to correlate strongly with 2-D mammographic percent density measurements ($r_s = 0.7$) of breast density (BD). Volpara is an automated mammographic density measurement software tool that measures BD volumetrically and therefore provides a better external standard to compare with UST's volumetric sound speed measurements. Since increased BD is known to increase the risk of developing breast cancer, sound speed images can potentially offer new insight into measurements of breast tissue without the use of ionizing radiation. The purpose of this work is to provide a direct comparison of UST vs Volpara measurements to determine the viability of sound speed as an imaging biomarker of BD.

Method and materials

A group of 100 women with benign and malignant findings in their breast underwent both a UST breast scan and had a Volpara reading of a mammogram at our local cancer center. Only those patients that received a UST scan within one year relative to the Volpara mammogram reading were selected. The UST scans occurred over a period ranging from May 2014 to February 2016 as Volpara was only used at our center during 2015. Spearman correlation coefficients were calculated to determine the strength of the correlations. between the Volpara and UST measurements of BD.

Results

The results show a very strong correlation ($r_s = 0.85$) between Volpara volumetric percent density and UST whole breast sound speed values. This correlation is significantly stronger than those from previous 2-D studies ($r_s = 0.85$ vs $r_s = 0.7$, respectively). The strong correlation suggests that UST sound speed is a viable imaging biomarker for measuring BD.

Conclusion
A comparison between UST and Volpara volumetric measurements shows a stronger correlation than those found in earlier studies that used 2-D mammographic BD measurements. This result strengthens the potential role of UST as an alternative biomarker of BD.

Clinical relevance / application
Elevated BD strongly increases a woman's risk of developing breast cancer. Since UST is non-ionizing, BD could be studied in a broader population of women, including those below screening age. UST provides quantitative information obtained without compression and radiation that has the potential to provide more accurate BD information, leading to better stratification of breast cancer risk.

Keywords: Breast Density, Breast Soundspeed, Volpara

1 Objectives

Increased mammographic breast density (BD) has been linked to increased risk of developing breast cancer [1,2]. Breast density has traditionally been evalutated using 2-D mammographic images. It stands to reason that the entire volume of dense tissue should offer more insight into the risk of developing breast cancer compared to the projected area of the dense tissue of a compressed breast that is visible in mammography. Volpara is an automated mammographic density measurement software tool that measures BD from mammograms using volumetric estimates [3]. Volpara BD measures are given as both a conventional letter grade and as a quantitative percent density score.

Ultrasound tomography (UST) can produce multiple high quality images of breast tissue [4,5]. One of these image types, soundspeed imaging, can be used to directly map breast tissue density. The volumetric nature of UST imaging allows for uncompressed images of the breast to be created without the use of ionizing radiation. Quantitative volume averaged UST sound speed measurements made from have been shown to correlate strongly with 2-D mammographic BD measurements ($r_s = 0.7$) [6–9]. These results were accomplished using a ray-based reconstruction algorithm which resulted in low image quality. Recent improvements in UST waveform reconstructions have allowed for high resolution soundspeed maps of breast tissue which could potentially offer new insight into measurements of breast tissue (Figure 1) [10]. The aim of this work is to provide a direct comparison of UST vs Volpara measurements to determine the viability of sound speed as an imaging biomarker of BD.

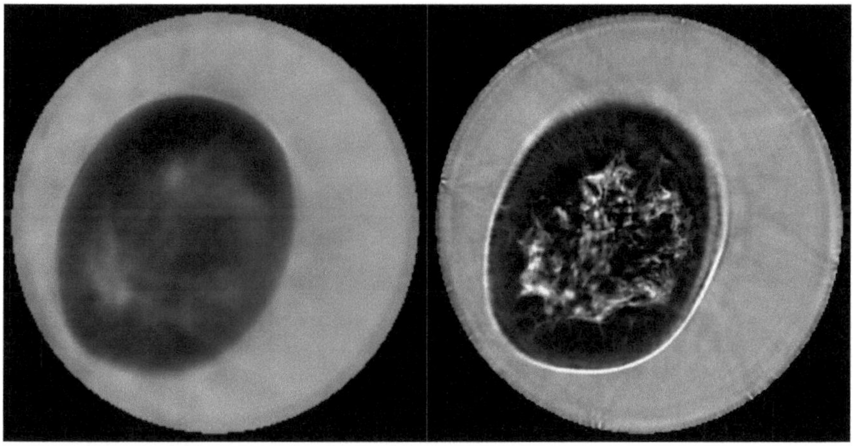

Figure 1: Comparison of the ray based soundspeed reconstruction (Left) with the waveform based soundspeed reconstruction (Right) of the same scan data. Note the dramatic increase in image quality that the waveform image now contains.

2 Methods

2.1 Patient Recruitment

A group of 100 women with benign and malignant findings in their breast underwent both a SoftVue UST breast scan and had a Volpara reading of a mammogram at our local cancer center, the Karmanos Cancer Institute (KCI). Only those patients that received a UST scan within one year relative to the Volpara mammogram reading were selected. The UST scans occurred over a period ranging from May 2014 to February 2016 as Volpara was only used at our center during 2015. Mammograms were performed on both breasts whereas UST scans were usually performed on one breast with a potentially suspicious finding present.

2.2 UST Soundspeed Images

The raw UST data that was collected at the time of the scan was used to reconstruct waveform soundspeed images using an up to date reconstruction algorithm. Whole breast volume averaged soundspeed (VASS) was then measured from the soundspeed image and used as the UST measure of BD. Breast tissue was segmented from the background water bath using masks created from an automated algorithm that is based on the SoftVue Wafer images. The Wafer images naturally provide enhanced contrast between the breast tissue and the background water bath compared to the soundspeed images where the soundspeed of water is intermediate to that of the range of breast tissue. The voxels remaining in the masked soundspeed images can then be averaged to compute the VASS value for the case and counted to compute the total breast volume.

A k-means segmentation algorithm was then run on the soundspeed images to separate the tissue into dense and non-dense regions. Segmentation of the tissue into its subregions allowed for calculation of the percent of the volume that was dense in a manner similar to that used for mammography. Additionally, the quantitative nature of the UST soundspeed images also allowed for the calculation of the mean values of these subregions.

2.3 Volpara Measurements and Statistical Analysis

As part of their screening mammogram, a Volpara reading was also performed. As part of the Volpara report, the total volume, dense volume, percent density (PD) and density grade were all recorded for the same breast that was scanned using UST. Spearman correlation coefficients were run to compare the corresponding UST and mammographic breast density measures

3 Results

On average, the UST scan occurred 35 days before the corresponding mammogram. At the time of the UST scan, the average age, weight, height and BMI of the patient population scanned was 51.5 years, 178.9 lbs, 64.7 inches and 30.0 kg/m^2. The patient population consisted of 75% African American, 18.5% White and 6.5% other races.

The total volume of breast tissue by UST was approximately 10% lower than that measured by Volpara. However, UST measured a greater volume of dense tissue than Volpara measured. These combined effects resulted in the UST PD measurements being much higher than the Volpara PD measurements. All comparable measures were tested using a paired t-test and all showed a p value of p ¡ 0.001. Results are summarized in Table 1.

Volume Measure	Volpara Average	UST Average
Total Breast Volume (cm^3)	1103	1003
Dense Breast Volume (cm^3)	80	191
Fatty Breast Volume (cm^3)	1023	812
Percent Density (%)	8.8	22.6

Table 1: Comparison of Volpara and UST Breast Volume Measurements

Very strong correlations were also observed between the UST VASS measurement and the Volpara volumetric PD (r_s=0.85). A slightly stronger correlation was also observed between

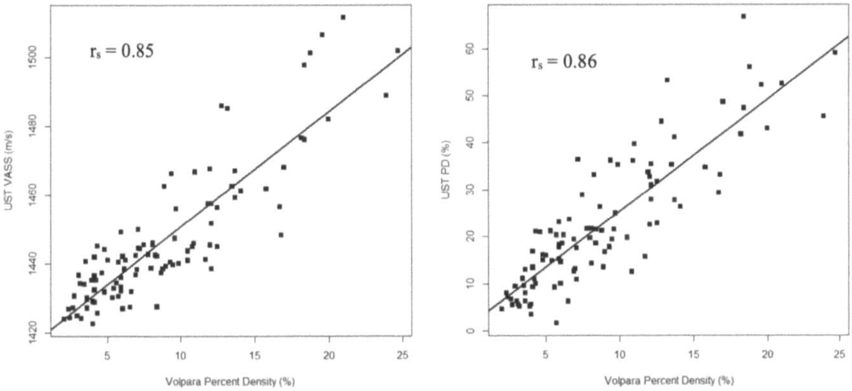

Figure 2: (Left) Plot of the UST VASS for each case verse the Volpara PD measurement. (Right) Plot of the UST PD measurement for each case vers the Volpara PD measurement.

the two comparable PD measurements made by both UST and Volpara (r_s=0.86). These correlations are plotted in Figure 2.

Boxplots of the UST VASS, UST PD and Volpara PD measurement grouped by the Volpara density grade were also recorded and can be seen in Figure 3. The plots show that the Volpara PD is separated well into the different categories, but some overlap between categories can be seen for the UST measurements.

A frequency distribution of the mean sound speed of the dense and fatty regions can be seen in Figure 4. This figure shows that the dense and fatty regions created by the k-means clustering algorithm are generally well separated but that there is still a range of possible sound speed values that these subregions can take on. Fatty tissue also appears to be more strongly peaked around the lower possible soundspeed values whereas dense tissue appears to spread out over a wider range.

4 Discussion

The volumes of tissue measured by Volpara and UST differed slightly and can likely be explained by the differences in the imaging modalities. Although UST is volumetric in nature, the total volume measured underestimated that measured by Volpara. The differences may be due to insufficient coverage of the breast tissue near the chest wall where some UST scans did not sufficiently reach far enough back. However, this overestimation of total breast volume by Volpara has also been shown to occur in when compared to MRI, a modality that images the breast in a similar fashion to UST [11].

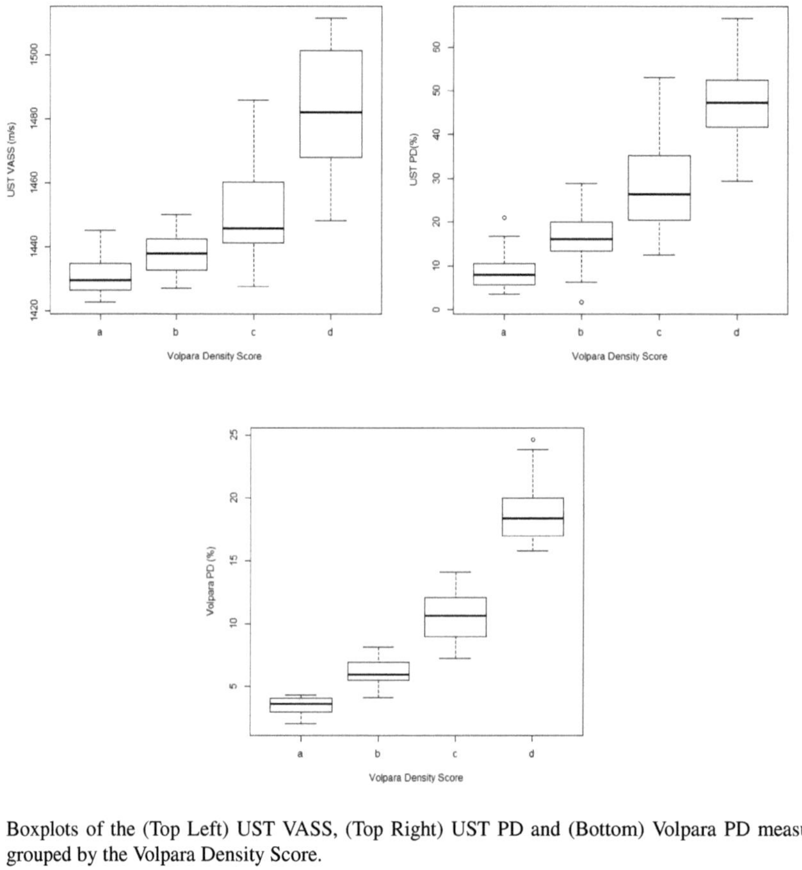

Figure 3: Boxplots of the (Top Left) UST VASS, (Top Right) UST PD and (Bottom) Volpara PD measuremens grouped by the Volpara Density Score.

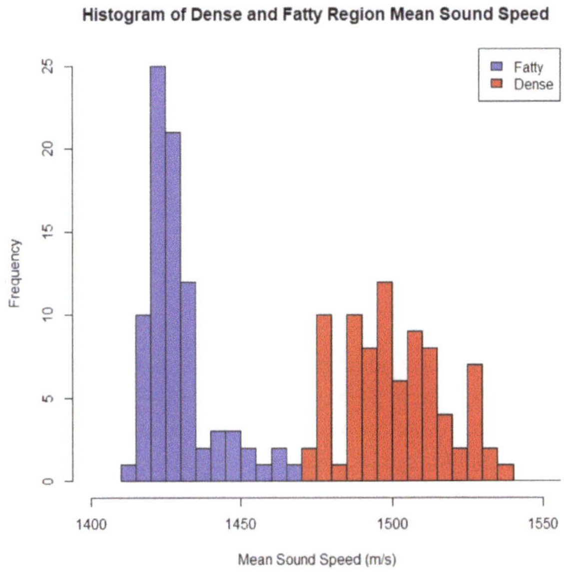

Figure 4: Distribution of the mean soundspeed of the dense and fatty subregions for each soundspeed image.

UST does not require compression of the breast to image it, so the distribution of dense tissue throughout the breast is unlikely to be obscured or distorted. The lack of overlapping tissue likely explains why the measurement of dense volume was much higher for UST compared to Volpara. Since the breast is imaged in a more natural position in UST, the dense tissue is much more likely to be completely accounted for.

The correlations between Volpara and the waveform density measures (Figure 2) were stronger than previous correlations made with ray based UST reconstructions and 2-D mammographic density measures [8,9]. This indicates that volumetric estimates likely produce more information than those based on straightforward 2-D measures.

Since the classification of the Volpara density grade is linked to the calculation of the Volpara PD measurement, it is expected that there is little overlap between the two groups as seen in Figure 3. When grouped by the Volpara density grade, the UST density measures show a wide range of possible values, especially for those patients with a grade of "c" and classified as "dense". This indicates that UST density measures may provide a finer grade of density classification on patients with a higher risk of developing breast cancer. UST VASS has already been shown to potentially offer stronger density based breast cancer risk assessment than mammography [12].

Finally, UST also offers additional information regarding the two major subregions, dense and fatty tissue, that comprise breast tissue. This is information unique to UST and can be seen in Figure 4 which shows how the average soundspeed of the two regions can vary. Fatty tissues do appear to be more uniform throughout showing a large peak in the histogram at the lower soundspeed values. However, the dense tissue does appear to be more distributed over a larger range of values. Dense and fatty tissue are not similar across patients and calculating density as simply the relative amounts of these two tissue may not accurately describe the conditions within the breast. It is unknown how the distribution of the soundspeed values in the subregions will affect breast cancer risk though.

5 Conclusions

Elevated BD strongly increases a woman's risk of developing breast cancer. Since UST is non-ionizing, BD could be studied in a broader population of women, including those below screening age. UST provides quantitative information obtained without compression and radiation that has the potential to provide more accurate BD information, potentially leading to better stratification of breast cancer risk.

References

[1] V. A. McCormack and I. dos Santos Silva, Breast Density and Parenchymal Patterns as Markers of Breast Cancer Risk: A Meta-analysis, Cancer Epidemiol. Biomarkers Prev., vol. 15, no. 6, pp. 1159–1169, 2006, doi: 10.1158/1055-9965.epi-06-0034.

[2] N. F. Boyd, H. M. Jensen, G. Cooke, and H. L. Han, Relationship Between Mammographic and Histological Risk Factors for Breast Cancer, JNCI J. Natl. Cancer Inst., vol. 84, no. 15, pp. 1170–1179, Aug. 1992, doi: 10.1093/jnci/84.15.1170.

[3] A. Gubern-Mérida, M. Kallenberg, B. Platel, R. M. Mann, R. Martí, and N. Karssemeijer, Volumetric Breast Density Estimation from Full-Field Digital Mammograms: A Validation Study, PLoS One, vol. 9, no. 1, p. e85952, Jan. 2014.

[4] N. Duric et al., Detection of breast cancer with ultrasound tomography: First results with the Computed Ultrasound Risk Evaluation (CURE) prototype, Med. Phys., vol. 34, no. 2, pp. 773–785, 2007.

[5] N. Duric et al., Breast Imaging with the SoftVue Imaging system: First results, in Medical Imaging 2013: Ultrasonic Imaging, Tomography, and Therapy, 2013, pp. 8619–8675.

[6] C. Glide, N. Duric, and P. Littrup, Novel approach to evaluating breast density utilizing ultrasound tomography, Med. Phys., vol. 34, no. 2, pp. 744–753, 2007.

[7] C. K. Glide-Hurst, N. Duric, and P. Littrup, Volumetric breast density evaluation from ultrasound tomography images, Med. Phys., vol. 35, no. 9, pp. 3988–3997, 2008.

[8] N. Duric et al., Breast density measurements with ultrasound tomography: A comparison with film and digital mammography, Med. Phys., vol. 40, no. 1, pp. 13501–13512, 2013.

[9] M. Sak et al., Using Speed of Sound Imaging to Characterize Breast Density, Ultrasound Med. Biol., vol. 43, no. 1, pp. 91–103, Jan. 2017, doi: 10.1016/J.ULTRASMEDBIO.2016.08.021.
[10] G. Y. Sandhu, C. Li, O. Roy, S. Schmidt, and N. Duric, Frequency domain ultrasound waveform tomography: breast imaging using a ring transducer, Phys. Med. Biol., vol. 60, no. 14, pp. 5381–5398, 2015, doi: 10.1088/0031-9155/60/14/5381.
[11] E. A. M. O'Flynn et al., Ultrasound Tomography Evaluation of Breast Density: A Comparison With Noncontrast Magnetic Resonance Imaging, Invest. Radiol., vol. 52, no. 6, pp. 343–348, Jun. 2017, doi: 10.1097/RLI.0000000000000347.
[12] N. Duric et al., Using Whole Breast Ultrasound Tomography to Improve Breast Cancer Risk Assessment: A Novel Risk Factor Based on the Quantitative Tissue Property of Sound Speed, J. Clin. Med., vol. 9, no. 2, p. 367, Jan. 2020, doi: 10.3390/jcm9020367.

Using Ultrasound Tomography to Monitor Response to Neoadjuvant Chemotherapy

C. Li[1], M. Sak[1], D. Chen[1,2], N. Duric[1,2], P. Littrup[1], and R. Brem[3]

[1] *Delphinus Medical Technologies, Inc, Detroit, MI, USA, E-mail: cli@delphinusmt.com*
[2] *Karmanos Cancer Institute, Detroit, MI, USA*
[3] *George Washington University School of Medicine & Health Sciences, D.C., USA*

Abstract

Our previous studies have shown that tissue sound speed, derived from ultrasound tomography (UST) measurements, is an imaging biomarker that can be used to characterize breast tissue. Since ultrasound is relatively benign and inexpensive, the purpose of this study was to evaluate sound speed as a reliable and cost-effective imaging biomarker for assessing neoadjuvant chemotherapy (NAC).

Fourteen patients undergoing neo-adjuvant chemotherapy for invasive breast cancer, were serially examined with UST throughout their treatment. The two parameters measured were the volume and the sound speed of the tumor. Response curves of both parameters were plotted for each study participant. Pathology results were used to classify participants as complete or partial responders based on whether they achieved pathologic complete response (pCR) or not.

In the partial response group, volume and sound speed showed uncorrelated and sometimes opposing changes with time while the complete response group showed steep changes in time with simultaneous drops in volume and sound speed. Furthermore, large drops in volume and sound speed in the first 3 weeks of treatment appeared to be predictive of pCR.

UST has potential for non-invasive, rapid identification of partial vs complete responders in women undergoing NAC without the use of either a radiotracer or gadolinium. Clinical decision making would improve by transitioning non-responders to alternative treatment quickly and by demonstrating effective response to NAC. A future larger study will test the predictive power of UST prospectively.

Keywords: Ultrasound tomography, neoadjuvant chemotherapy, sound speed

1 Introduction

Locally advanced breast cancer represents a difficult clinical problem. Many patients with locally advanced disease experience relapse and eventual death from the disease [1]. Data from the National Cancer Institute's Surveillance, Epidemiology, and End Results (SEER) program indicate that approximately 14,000 women a year are diagnosed with locally advanced breast cancer [2]. The 5-year relative survival rate for women with stage III breast cancer is about 55%. Neoadjuvant chemotherapy (NAC) increases the ability to control locally advanced breast carcinomas and promotes breast-conserving surgery [1-6]. However, not all patients respond to chemotherapy; if they do, their responses are highly variable. Patients who do not respond are less likely to benefit from NAC. Thus, identifying these patients earlier would allow a timely switch to a different regimen and/or would advance surgery. Patients in these categories would benefit by stabilizing and/or potentially reversing their disease, thereby reducing morbidity and mortality rates. [1-6] Clinically, there has not been a universal, cost-effective adoption of any technology or technique that helps accurately assess, monitor and predict individual patient response to NAC [7].

MRI and PET imaging have been shown to predict response as early as two weeks after treatment begins.[8,9]. Both imaging biomarkers have allowed correlation with the surgical pathology findings to assess concordance and enhance the potential for pre-operative planning. Unfortunately, the high costs of imaging associated with both MRI and PET have impeded research needed to verify outcomes and widespread acceptance of these imaging modalities. Thus, under current standard of care guidelines, oncologists and patients may not know how well a treatment is working until well into the course of treatment. The ability to identify non-responders early in the treatment process would provide potentially crucial guidance for changing to alternative regimens thereby minimizing patient suffering from unnecessary NAC side-effects and preventing further tumor progression. Furthermore, predicting pCR would be highly beneficial for breast cancer drug development given the FDA's acceptance of pCR as an endpoint to support accelerated approval. In the absence of a practical method for monitoring response, significant improvements in image-assisted chemotherapy are unlikely.

UST is an emerging modality that uses ultrasound pulses in combination with tomographic inversion techniques to generate images of the entire breast volume [10-18]. In this feasibility study, UST is presented as a safe, cost-effective and comfortable imaging strategy to measure locally advanced breast tumor response to neoadjuvant chemotherapy (NAC), to predict clinical and pathologic response (pCR) early in the treatment process, building on a previous case study [19]. This study is enabled by the safe (radiation free), fast, repeatable and frequent measurements that provide a practical low-cost method for informing clinical decision making.

2 Methods

Fourteen patients were scanned with a UST prototype, on days when they received chemotherapy. All scans were performed at the Karmanos Cancer Institute (KCI), Detroit, MI, USA. Patients were recruited with Wayne State University's, IRB approval and the study was HIPPA compliant. The prototype was developed by members of our team and its principles of operation have been described in detail [16,18]. This feasibility study was conducted to test the hypothesis that tumor sound speed and volume are quantitative biomarkers to predict response soon after the start of NAC. The study was motivated by the observations that tumor size changes are observed in standard imaging, tumor softening can be deduced from manual palpation and tumor sound speed (a measure of tissue density) and volume are accurately measured and quantified with UST. The last observation provided a basis for quantifying changes in tumor volume and density. UST breast imaging was performed at each chemotherapy visit (weekly or biweekly depending on treatment regimen). Each UST exam yielded a sound speed image stack representing the 3-D volume of the breast. Tumors were segmented in each image slice and the slices combined to create a 3-D tumor image. Volume was then directly measured from the 3-D rendering. Volume averaged sound speed was calculated from the same rendering and used as a marker of tumor density. The process was repeated at each visit, yielding tumor volume and sound speed at multiple time points for each patient.

3 Results

The tumor volume and sound speed were successfully measured and recorded for each of the fourteen patients at weekly or biweekly intervals. The corresponding plots of the tumor response are shown in Figure 1, for the partial responders, and in Figure 2, for the complete responders (those that achieved pCR). Since the goal of the study was to examine early response, we limited the plotted data to the first 60 days post-initiation of NAC. Exponential curves, fit to the data, are also shown, to facilitate the interpretation of the plots. Decay times (1/e - characteristic times to change from baseline value), obtained from the curve fitting are also shown. The plots and fitting summaries are shown separately for the tumor volume and sound speed changes.

Since all data shown are normalized to their baseline values, all plots show initial values of volume and sound speed to be one. All subsequent data points are therefore relative numbers that can compared against the baseline value to characterize tumor changes. The exponential curves are presented in order to visualize the trends in the data. In the partial response group, volume and sound speed showed uncorrelated and sometimes opposing changes with time. Tumor volumes and sound speed changes can be declining, increasing or flat. In contrast, the complete response group shows steep changes in time with simultaneous drops in volume and sound speed.

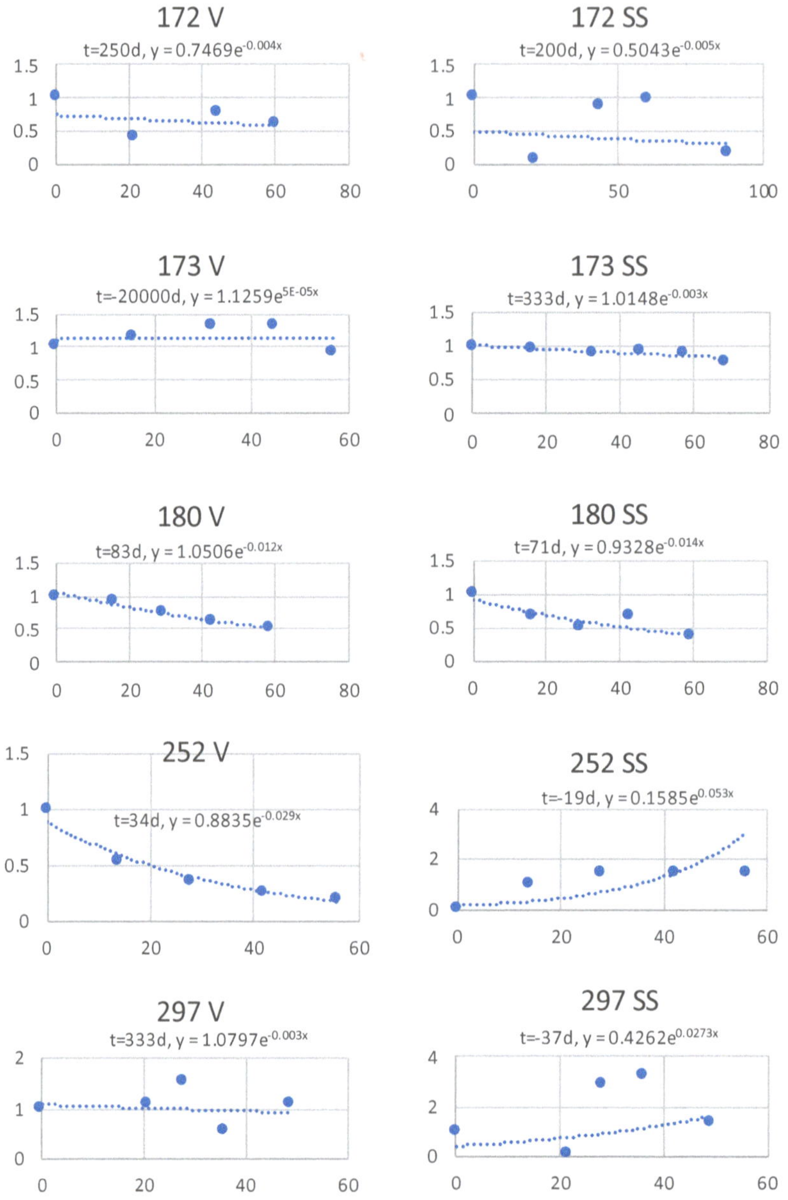
172 V
t=250d, y = 0.7469e^-0.004x
1.5
1
0.5
0
0 20 40 60 80
172 SS
t=200d, y = 0.5043e^-0.005x
1.5
1
0.5
0
0 50 100
173 V
t=-20000d, y = 1.1259e^5E-05x
1.5
1
0.5
0
0 20 40 60
173 SS
t=333d, y = 1.0148e^-0.003x
1.5
1
0.5
0
0 20 40 60 80
180 V
t=83d, y = 1.0506e^-0.012x
1.5
1
0.5
0
0 20 40 60 80
180 SS
t=71d, y = 0.9328e^-0.014x
1.5
1
0.5
0
0 20 40 60 80
252 V
t=34d, y = 0.8835e^-0.029x
1.5
1
0.5
0
0 20 40 60
252 SS
t=-19d, y = 0.1585e^0.053x
4
2
0
0 20 40 60
297 V
t=333d, y = 1.0797e^-0.003x
2
1
0
0 20 40 60
297 SS
t=-37d, y = 0.4262e^0.0273x
4
2
0
0 20 40 60

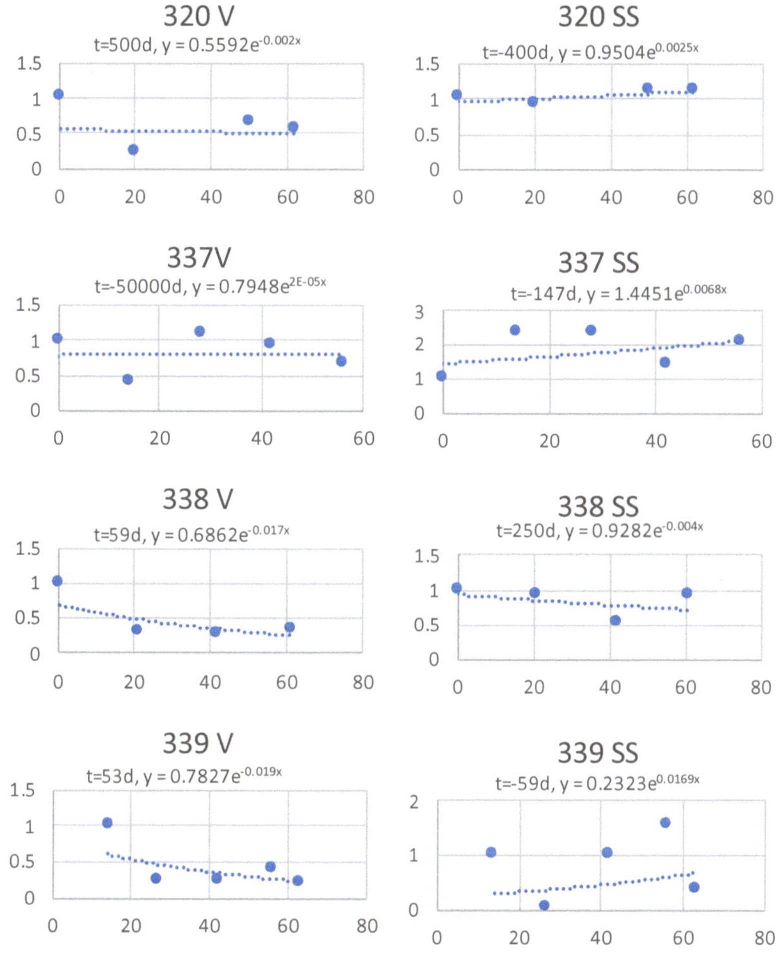

Figure 1: Volume (V) and sound speed (SS) responses for the first 60 days for the partial responders. Trendlines were fit using exponentials. The decay times are shown for each patient.

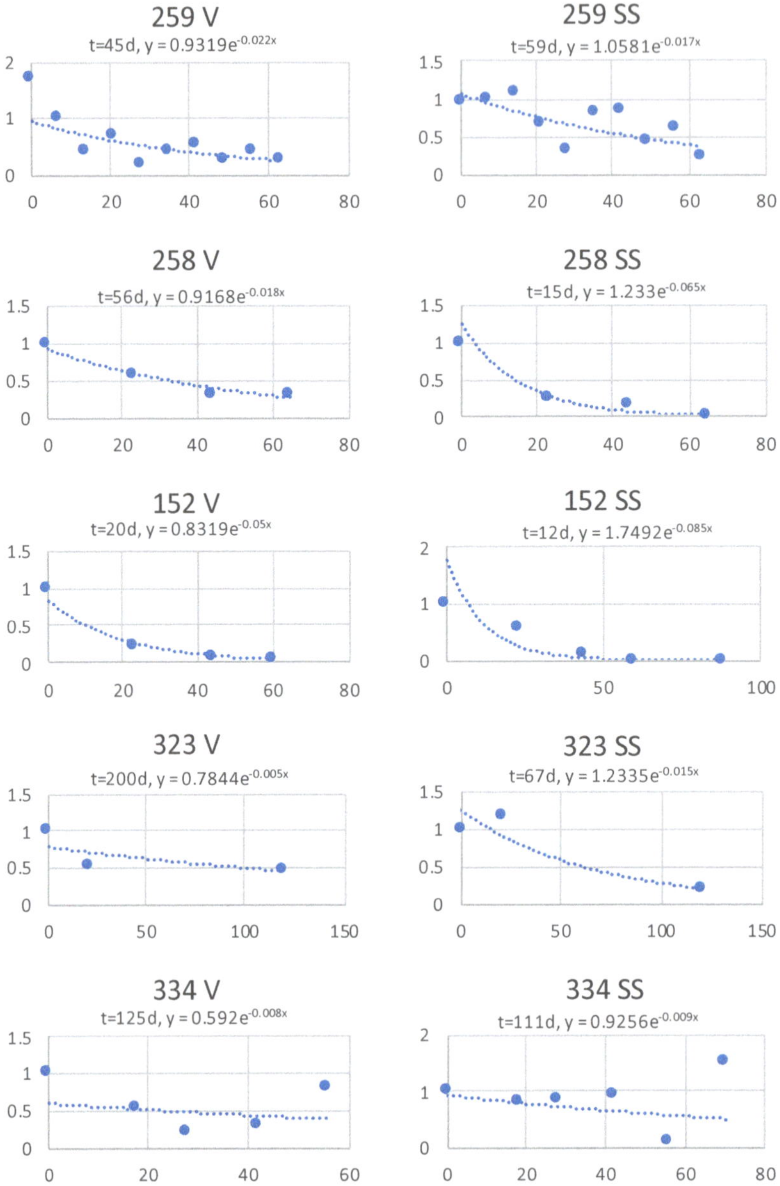

Figure 2: Volume (V) and sound speed (SS) responses for the first 60 days for the complete responders. Trendlines were fit using exponentials. The decay times are shown for each patient.

4 Discussion

The empirically observed changes in tumor sound speed are likely to be associated with the changing biomechanical properties of the tumor in its response to the chemotherapy. It has been known for some time that the speed of sound, c, and tissue density, ρ, are linearly correlated, such that c (km/s) = 1.12 ρ (g/cm3) + 0.39. Thus, UST generated sound speed images can be used to assess breast tissue density [20-21]. It is therefore biologically plausible that measured changes in sound speed may be driven by changes in tumor density which would be analogous to tumor assessment by palpation.

Tumor volume changes. In the partial response group, five out of the 9 cases show no decrease in tumor volume. In two cases, the volume of the tumor actually increases. Only four of the 9 case show a decline in tumor volume. In the complete response group, all five tumors show a volume decline. These results suggest that, statistically, patient groups that exhibit a tumor volume decline in the first 60 days are more likely to achieve pCR and thereby a more favorable outcome from their treatment. Larger, future studies would be required to determine the degree to which the volume response predicts pCR.

Tumor sound speed changes. In the partial response group, four out of the 9 cases show no decrease in tumor sound speed. In three cases, the volume of the tumor actually increases. Only five of the 9 case show a decline in tumor sound speed. In the complete response group, all five tumors show a decline in sound speed. These results suggest that, statistically, patient groups that exhibit a tumor sound speed decline in the first 60 days are more likely to achieve pCR and thereby a more favorable outcome from their treatment. Larger, future studies would be required to determine the degree to which the sound speed response predicts pCR.

Combined parameter changes. The volume and sound speed changes are not generally correlated, suggesting that they are independent parameters for assessing tumor response. For example, only 1/3 of the patients in the partial response group exhibited simultaneous declines in both volume and sound speed. In the complete response group, all five patients exhibited simultaneous declines in these parameters.

Additional examination of the above results shows that the complete response group often exhibited significant drops in volume and sound speed within the first 20 days. Observing such a drop early in the course of treatment could potentially be used to predict who will achieve pCR and who won't a after just 2 or 3 cycles of chemotherapy. Similarly identifying non-responders early would provide valuable guidance to the oncologist for either altering the treatment regimen or moving up the time of surgery.

Ongoing analysis of the numerical data generated by this study will quantify decline times and attempt to create cut points for separating the partial vs complete responders. These retrospectively determined cut points would be a first step in developing a clinical model that would predict prospectively the likelihood that an individual will achieve pCR early in the course of treatment.

While the sample of patients is small, our results provide some useful guidelines for future studies. Such studies could examine the predictive power of the combined volume and sound speed parameters with a prospective test of cut points defined either from the data generated in this study our as part of a future study. Furthermore, other UST parameters such as attenuation and stiffness could also be explored.

5 Conclusions

Our study demonstrates that UST can be used to monitor NAC and that the partial vs complete responders could be separated based on how tumor volume and sound speed change with time. UST has potential for non-invasive, rapid identification of partial vs complete responders in women undergoing NAC without the use of either a radiotracer or gadolinium. Larger studies are required to assess the predictive power of this type of tumor response.

References

[1] Low JA, Berman AW, Steinberg SM, Danforth DN, Lippman ME, Swain SM. Long-term follow-up for locally advanced and inflammatory breast cancer patients treated with multimodality therapy. J Clin Oncol 2004 Oct 22:4067-74. PMID: 15483018 [PubMed - indexed for MEDLINE]

[2] SEER website. http://seer.cancer.gov.

[3] Guarneri V, Broglio K, Kau SW, et al. Prognostic value of pathologic complete response after primary chemotherapy in relation to hormone receptor status and other factors. J Clin Oncol 2006 Mar 1;24(7):1037–44.

[4] Connor CS, Kimler BF, Mammen JM, McGinness MK, Wagner JL, Alsop SM, Ward C, Fabian CJ, Khan QJ, Sharma P. Impact of neoadjuvant chemotherapy on axillary nodal involvement in patients with clinically node negative triple negative breast cancer. J Surg Oncol 2014 Sep 29. doi: 10.1002/jso.23790. [Epub ahead of print].

[5] Tewari M, Krishnamurthy A, Shukla HS. Predictive markers of response to neoadjuvant chemotherapy in breast cancer. Surg Oncol. 2008 Dec;17(4): 301-11. doi: 10.1016/j.suronc.2008.03.003. Epub 2008 May 7.

[6] Eniu, A. "Can we tailor neoadjuvant (primary) systemic therapy for breast cancer?." memo-Magazine of European Medical Oncology 1.3 (2008): 143-147.

[7] Schegerin M, Tosteson AN, Kaufman PA, Paulsen KD, Pogue BW. Prognostic imaging in neoadjuvant chemotherapy of locally-advanced breast cancer should be cost-effective. Breast Cancer Res Treat. 2009 Apr;114(3):573-47. PMCID: PMC2807135

[8] Sharma U, Danishad KK, Seenu V, Jagannathan NR. Longitudinal study of the assessment by MRI and diffusion-weighted imaging of tumor response in patients with locally advanced breast cancer undergoing neoadjuvant chemotherapy. NMR Biomed, 2009 22(1): p. 104-13. doi: 10.1002/nbm.1245.
[9] Hatt, M., Groheux, D., Martineau, A., Espié, M., Hindié, E., Giacchetti, S., ... & Cheze-Le Rest, C. (2013). Comparison between 18F-FDG PET image–derived indices for early prediction of response to neoadjuvant chemotherapy in breast cancer. Journal of Nuclear Medicine, 54(3), 341-349.
[10] Greenleaf JF, Johnson SA, Bahn RC. "Quantitative cross-sectional imaging of ultrasound parameters." Ultrasonics Symposium, 1977 IEEE.
[11] Carson PL, Meyer CR, Scherzinger AL, Oughton TV. Breast imaging in coronal planes with simultaneous pulse echo and transmission ultrasound. Science, 1981. Dec 4;214(4525): 1141-3.
[12] Johnson, S A, et al. "From laboratory to clinical trials: An odyssey of ultrasound inverse scattering imaging for breast cancer diagnosis." J Acoust Soc America 2006 120(5): 3023-3023.
[13] Andre M P, et al. "High-speed data acquisition in a diffraction tomography system employing large-scale toroidal arrays." Int J Imaging Systems Technol 1997 8(1): 137-147.
[14] Liu DL, Waag RC. Harmonic amplitude distribution in a wideband ultrasonic wavefront after propagation through human abdominal wall and breast specimens. J Acoust Soc Am, 1997. Feb;101(2):1172-83.
[15] Marmarelis, VZ, Tae-Seong K, Ramez EN. "High-resolution ultrasound transmission tomography." Medical Imaging 2003. Int Soc Optics Photonics, 2003.
[16] Duric N, Littrup P, Poulo L, Babkin A, Pevzner R, Holsapple E, Rama O, Glide C. Detection of breast cancer with ultrasound tomography: First results with the Computed Ultrasound Risk Evaluation (CURE) prototype. Med Phys, 2007. Feb;34(2):773-85.
[17] N. V. Ruiter, G. Göbel, L. Berger, M. Zapf and H. Gemmeke, "Realization of an optimized 3D USCT", Proc. SPIE 7968, 796805 (2011)
[18] Ranger B, Littrup P, Duric N, Chandiwala-Mody P, Li C, Schmidt S, Lupinacci J. Breast ultrasound tomography versus MRI for clinical display of anatomy and tumor rendering; preliminary results. AJR Am J Roentgenol. 2012 Jan;198(1):233-9. doi: 10.2214/AJR11.6910. PMID: 22194502.
[19] Jessica Lupinacci; Neb Duric ; Peter Littrup ; Ding Wang ; Cuiping Li ; Steven Schmidt; Bryan Ranger ; Erik West ; Amy Szczepanski ; Olsi Rama ; Lisa Bey-Knight ; Lukasz Myc. Monitoring breast masses with ultrasound tomography in patients undergoing neoadjuvant chemotherapy ; Proc. SPIE 7629, Medical Imaging 2010: Ultrasonic Imaging, Tomography, and Therapy, 76291C (March 12, 2010); doi:10.1117/12.845647
[20] Jeong JW, Shin DC, Do SH, et al. Differentiation of cancerous lesions in excised human breast specimens using multiband attenuation profiles from ultrasonic transmission tomography. J Ultrasound Med. 2008 Mar; 27(3):435-51.
[21] Mast, T. D. (2000). Empirical relationships between acoustic parameters in human soft tissues. Acoustics Research Letters Online, 1(2), 37-42.

Parallel Optimization for Ultrasound Computed Tomography using GPU

Y. Zhou[1], H. Hou[3], S. Wang[4], Q. Zhou[2], M. Yuchi[1,2], M. Ding[1,2], and Y. Zheng[1]

[1] *WuHan WeSee Medical Imaging Co. Ltd., China, E-mail: m.yuchi@hust.edu.cn*
[2] *School of Life Science and Technology, Huazhong University of Science and Technology, Wuhan, Hubei, China*
[3] *Cluster and Grid Computing Lab, Huazhong University of Science and Technology, Wuhan, Hubei, China*
[4] *Hubei University of Technology, Wuhan, Hubei, China*

Abstract

Background
GPU has been widely used in imaging and image processing because of its powerful parallel computing ability. The application of GPU in ultrasound computer tomography (USCT) has attracted many researchers in recent years. The USCT system (Lucid) developed in the Medical Ultrasound Laboratory uses a ring transducer which contains 2048 elements to capture the radio-frequency (RF) data. The amount of the RF date for one slice can reach up to 30 Gbytes, which brings great difficulties for image reconstruction. This paper introduces the image reconstruction acceleration of USCT with GPU.

Method
The process for the image reconstruction on GPU is as follows. First, all RF data are captured and saved on the server. Then the RF data are transferred to GPU memory and filtered. Finally, the image is reconstructed based on synthetic aperture focusing technique (SAFT) algorithm in GPU. The total RF data can be regarded as a 3-dimensional matrix. Considering that the memory of GPU is limited, the data are distributed into blocks and transferred to the memory sequentially and processed parallelly in GPU. Each block contains a certain number (N) of element RF data. If N is too small, the frequency of the data swap between the host memory, which results in low efficiency usage of GPU. If N is too large, the GPU memory cannot store the required raw data, which makes the hit rate of GPU memory low. The optimization of N is the main object of this work.

Result
The experimental results show that N = 128 is the optimal solution in our system. The processing time reduced to 38 seconds for the 2048 x 2048 image reconstruction, comparing to 4 minutes 10 seconds without GPU acceleration.

Random Field Interferometry for Medical Ultrasound

I.E. Ulrich[1], C. Boehm[1], and A. Fichtner[1]

[1] *ETH Zurich, Zurich, Switzerland, E-mail: ines.ulrich@erdw.ethz.ch*

Abstract

Ultrasound computed tomography (USCT) is frequently used for medical purposes to image soft tissue body parts, as for instance the breast.

Breast cancer detection using USCT usually works with a collection of ultrasound scans that measure the pressure wavefield emitted by individual transducers. This often requires a large number of emitter-receiver pairs to obtain a good coverage of the domain of interest, and careful calibration of the emitting transducers using reference measurements in water.

We present a novel approach to obtain time-of-flight measurements between receiver pairs in an USCT setup by applying the interferometry principle. By substituting active emitters with virtual ones, specific source imprints are eliminated, thus avoiding the need for reference measurements and calibration. Using interferometry and virtual emitters, we retrieve Greens functions between any two measurement locations, which can be used as new data for the inverse problem. The proposed method gives new perspectives to shorten the acquisition time of an entire USCT data set as well as to increase the data coverage.

We perform numerical experiments using a phantom representing a 2D cross-section of a breast and include some variations in the speed of sound inside the phantom to model regions with malignant cells. Using cross-correlations and stacking of A-scans, we show that one can extract the travel time between any pair of transducer positions from random wavefields, created by a superposition of individual source wavefields. Those data are then used in a time-of-flight inversion to reconstruct the speed of sound of the breast phantom.

Attenuation Image Reconstruction for Ultrasound Computed Tomography using FBP algorithm

Y. Wu[1], X. Fang[1], J. Song[1], L. Zhou[1], Q. Zhang[1], Q. Zhou[1], K. Liu[1], Z. Quan[1], M. Ding[1,2], and M. Yuchi[1,2]

[1] *Huazhong University of Science and Technology, Wuhan, China*
[2] *WuHan WeSee Medical Imaging Co. Ltd., China*

Abstract

Background
Ultrasound Computed Tomography (USCT) is a promising medical imaging tool, which reconstructs the images of the object using both reflected and transmitted ultrasound signals. Attenuation tomography of USCT has been studied by many researchers since its great clinical potential of early breast cancer detection.

Material and Method
The Lucid USCT system developed in Medical Ultrasound Lab was used to capture the breast phantom signals. The system contains one 2048-element ring array with the center frequency 3.0 MHz. The diameter of the ring is 220 mm, and the sampling frequency is 25 MHz. Filtered back projection (FBP) is widely adopted as a classical analytic reconstruction method for image reconstruction. In this study, a fan-beam FBP algorithm is used for attenuation tomography of the breast phantom which has 3 cylinders with different attenuation coefficients to mimic the cyst, benign, and malignant tumors.

Results
The size of the reconstructed image is 1024x1024, and the 3 cylinders can be clearly shown and distinguished from the background. There exits circle artifacts around the cylinders.

Conclusion
The result shows that the fan-beam FBP algorithm can reconstruct the image of the breast phantom and distinguish the imaging objects with different attenuation coefficients. The artifacts in the reconstructed image are to be resolved in the future work.

Refraction corrected transmission imaging based on Bézier curves: first results with KIT 3D USCT

F. Zuch[1], T. Hopp[1], M. Zapf[1], and N.V. Ruiter[1]

[1] *Karlsruhe Institute of Technology, Karlsruhe, Germany, E-mail: franziska.zuch@kit.edu*

Abstract

The computational complexity of wave-based image ultrasound transmission reconstructions is very high especially in 3D. Hence, ray-based approximations are often used. Bent ray approaches like the fast marching method are time-consuming in comparison to the straight ray approximation. The Bézier curve technique for refraction approximation, introduced by Perez-Liva et al., could be a compromise between computational effort and image quality. In this work, the method was extended to 3D and an evaluation on two 3D datasets was carried out. The first dataset was simulated with the k-wave toolbox, with hemispherically arranged emitters and receivers and a sound speed inhomogeneity. With a mean squared error of 9.98 m/s the method produces similar results as the fast marching method (11.85 m/s). The object size is represented better in comparison to the straight ray method. Applied to real data of the KIT 3D USCT both methods performed similarly. The object size was better preserved using bent ray methods. The Bézier bent ray method shows first promising results in 3D as it leads to similar results than the fast marching method.

Keywords: USCT, transmission reconstruction, Bézier curve

1 Introduction

Ultrasound computer tomography is a new imaging modality for breast cancer detection [1]. Both transmission and reflection information can be obtained. Transmission images provide quantitative information about tissue properties like sound speed and attenuation [2, 3]. The focus of this paper is on the sound speed reconstruction as part of the transmission tomography.

For the image reconstruction, the wave propagation through tissue needs to be described (forward model) and an inverse problem needs to be solved for the image given the measured pressure fields and the forward model. The computational complexity of wave-based image reconstructions is very high, especially in 3D at high frequencies [4]. For this reason, approximations are required. Here, ray-based approaches are commonly used to describe the path of propagation between emitter and receiver. The straight ray approximation neglects physical effects like refraction. Bent ray approaches like the widely used fast marching method [5, 6] may include this effect yet are still computationally demanding [7], especially in 3D.

Pérez Liva et al. [3, 7] introduced a ray-based method where a single refraction is approximated with a quadratic Bézier curve. The idea is that including the refraction of the water-tissue interface leads already to an improvement. A brute force search for the fastest curve out of a set of Bézier curves with different curvature connecting emitter and receiver is applied. The presented method was developed for the application in 2D. In this paper it is expanded to 3D, for the use on KIT 3D USCT, and an evaluation for 3D datasets is carried out.

2 Methods

2.1 Bent ray calculation with Bézier polynom

The basic idea of Pérez Liva et al. [3] is to link two points, which in USCT are emitter and receiver, with a curve described by an Bézier polynomial [8]. For such a curve the sound speed map is used to derive the time of flight (TOF) of an ultrasound pulse travelling along this curve. Using a brute force search the curve with lowest TOF is determined.

For implementation in 3D, we consider a quadratic Bézier curve, discretized by t, which is described by the following equation:

$$C(t) = (1-t)^2 P_0 + 2t(1-t)P_1 + t^2 P_2, t \in [0,1] \,. \tag{1}$$

P_0, P_2 correspond to the start- and endpoint of the curve. In 3D USCT these positions are given by the emitter and receiver coordinates. P_1 is a free point. It controls the curvature of the Bézier curve. By moving this control point, different curves between emitter and receiver can be generated. The free points P_1 are distributed along and vertical to the direct line connecting emitter and receiver in order to derive a set of possible curves connecting P_0 and P_2. The distribution was calculated via linear equation from the emitter-receiver-centre with the given direction vectors (see Fig. 1 (a)):

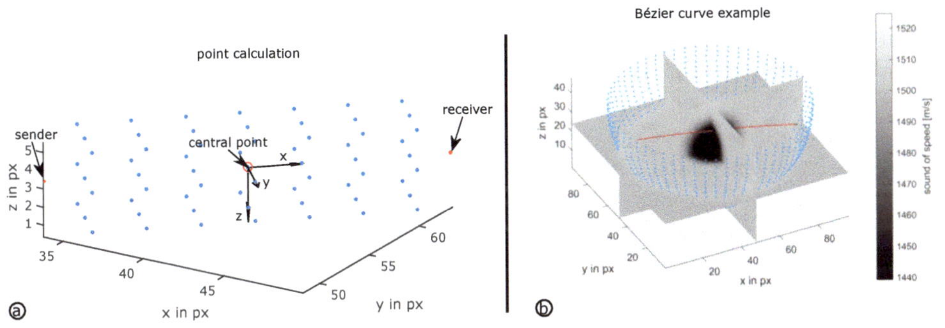

Figure 1: Bézier curve calculation. (a) shows the calculation of the 3D point distribution for determining the set of Bézier curves. It is calculated by the three direction vectors along and orthogonal to the emitter receiver line. The start and endpoint is given by the emitter and receiver ((b), blue points). An example Bézier curve (red) is shown on the underlying first guess sound of speed reconstruction.

$$\vec{v_x} = \overrightarrow{P_0P_2}$$
$$\vec{v_y} = \vec{v_x} \times \begin{pmatrix} 0 \\ 0 \\ 1 \end{pmatrix} \tag{2}$$
$$\vec{v_z} = \vec{v_x} \times \vec{v_y} \, .$$

The selection of the optimal curve is based on a sound speed map. As a first guess, the sound speed map reconstructed with straight rays is used. For each discretized point of the curve the corresponding sound speed value (c_t) is obtained via linear interpolation and the euclidean distance between the sample points is calculated to determine la_t in equation 3. The TOF of an ultrasound pulse travelling along each curve is then defined by:

$$TOF = \sum_{t=1}^{t=k} \left(\frac{1}{c_t} la_t\right) . \tag{3}$$

The curve with the lowest TOF is selected as the optimal one following Fermat's principle [9]. An example curve on an underlying sound speed map is shown in Fig. 1 (b).

2.2 Reconstruction

The image reconstruction is formulated as a linear equation system

$$\vec{b} = M\vec{x} \, . \tag{4}$$

For reconstructing the sound speed image $\vec{x}$, the equation system needs to be solved for $\vec{x}$ given the measured time of flight $\vec{b}$ and the system matrix M. It is solved iteratively by an algebraic reconstruction technique using total variation minimization (TVAL3 algorithm [10, 11]). The time of flight values $\vec{b} = (b_1, ..., b_m)^T$ are gained by calculating the time differences of the recorded signal between the patient measurement and an empty measurement of each emitter-receiver-combination. Matrix M contains the ray paths which describe the wave propagation. For each emitter-receiver combination the voxel discretized Bézier curve is recorded in M. This results in a matrix size of number of measurements $(m) \times$ number of voxels (n).

Considering the refraction, the reconstruction is an iterative procedure. The solution $\vec{x}$ is taken as the new sound speed map on which the Bézier curve method is applied on and M is updated accordingly. By solving the equation system a new estimate for $\vec{x}$ is generated. This proceeding is repeated until convergence (change between iterations $< \varepsilon$) or the number of maximum iterations is reached.

3 Results

3.1 2D Evaluations

Before applying the method to 3D, we analyzed the characteristics of the Bézier curve method in 2D experiments due to decreased computing time. The change of the mean squared error (MSE) over the iterations was analyzed. 20 iterations have been carried out. A simple experimental set up was used, with emitter and receiver arranged circular and an circular sound speed inhomogeneity in the centre (see Fig. 2). For the emitter-receiver-ring a radius of 5 cm has been chosen. 200 emitter and 200 receiver were placed. The center frequency was 2.5 MHz with a bandwidth of 1 MHz. The data had been simulated with k-wave [12]. For this setup the MSE decreases the most on the first iterations. For this reason, the following bent ray reconstructions in 3D have been calculated with one iteration as a trade-off between computing time and accuracy. The minimum MSE for the fast marching bent ray reconstruction was reached in the second iteration with 35,00 m/s, for Bézier it was the tenth iteration with 25,64 m/s (straight ray 74,45 m/s).

Furthermore an experiment with third order Bézier curves had been taken for one exemplary ray. The results have been compared to the path calculated using Fermat's principle for a simple setup (see Fig. 3 top) [13]. The ray of the cubic Bézier curve seems to get closer to the calculated one following Fermat‘s principle. The length deviation is reduced to 1.2 mm in comparison to 1.9 mm. However, the calculation time for the selected parameters increase around factor 100 as the number of free points doubles and all permutations have to be evaluated in the brute force search. Due to the high computational cost and the relatively low gain by cubic Bézier curves on simple object evaluation, we decided to carry out 3D

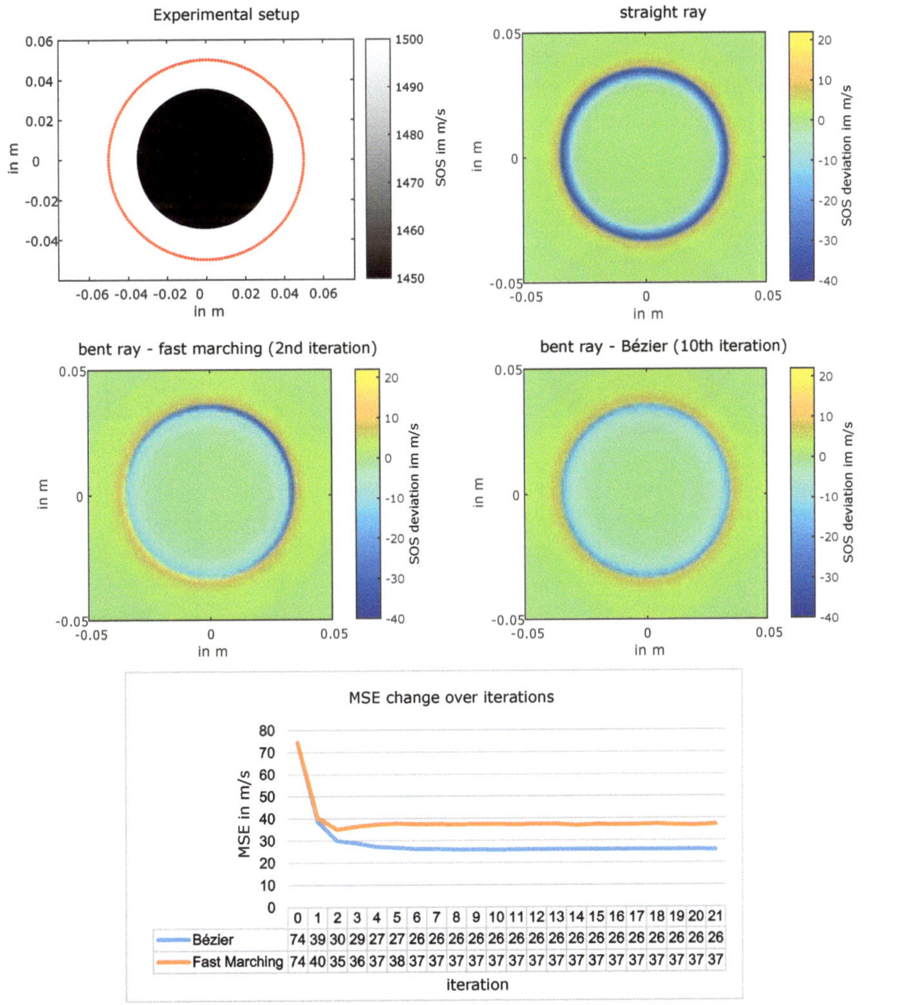

	0	1	2	3	4	5	6	7	8	9	10	11	12	13	14	15	16	17	18	19	20	21
Bézier	74	39	30	29	27	27	26	26	26	26	26	26	26	26	26	26	26	26	26	26	26	26
Fast Marching	74	40	35	36	37	38	37	37	37	37	37	37	37	37	37	37	37	37	37	37	37	37

Figure 2: Evaluation of MSE change over iterations. On the upper left the experimental setup is presented, with the emitter/receiver arrangement collar coded red. At the bottom the MSE change over the iterations for Bézier and fast marching is shown. The difference images of zero iteration (straight ray) and the bent ray reconstructions to the reference image are shown.

reconstruction with quadratic Bézier curves. Cubic Bézier curves may be investigated further in future, when a performant GPU implementation is available.

Furthermore the discretization of the Bézier curve, i.e. the step size for t in equation 1 was calculated. The discretization is coupled to the distance in pixels between emitter and receiver. The number of discrete points on the Bézier curve is given by this distance times a factor. Due to the curvature of the Bézier curve, equidistant distribution of t leads to an irregular point distribution in space, we investigated which factor is required such that the largest distance between two discrete points on the Bézier curve is smaller than half the pixel size in order to fulfill the sampling theorem. With the same experimental 2D setup as before we found that the factor needs to be equal or higher than 3.6.

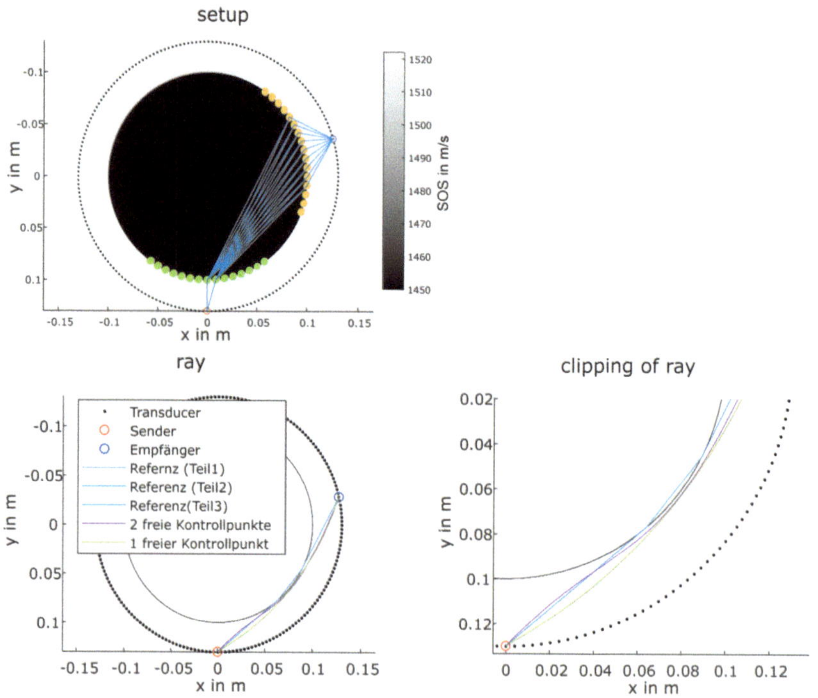

Figure 3: Evaluation of quadratic and cubic Bézier curves. bent rays according to Fermat's principle. Reference beams are calculated using Fermat's principle (top image). The black dots represent the transducers. The example receiver is outlined in blue, the example transmitter is outlined in red. Green or yellow are the discretized points on the circular surface. In blue exemplary paths are given describing possible paths from transmitter to receiver. At the bottom the quadratic and cubic Bézier curves are shown in comparison to the calculated one, with a detailed view on the bottom right.

3.2 Application in 3D

A comparison between Bézier bent ray reconstruction, fast marching bent ray reconstruction and straight ray reconstruction was carried out. For the bent ray reconstruction, one iteration was performed based on the straight ray reconstruction. For the 3D evaluation two datasets are used, one with simulated data and one recorded dataset with KIT 3D USCT [14]. For simulation a simple setup with one sound speed inhomogeneity was selected to examine the main refraction of the water object/tissue interface. In addition more complex phantom data from KIT 3D USCT have been used.

3.2.1 3D k-wave simulation data

For evaluation in 3D simulated A-scan data was used first. We applied a k-wave [12] simulation with a setup of 231 emitter and 1781 receiver. Emitter and receiver are arranged in an hemisphere around a spherical object (0.58 cm diameter) with sound speed inhomogeneity (background: 1500 m/s, object: 1450 m/s). Due to computing time and memory requirements the aperture size with a radius of 1 cm and a depth of 1 cm is smaller than the KIT 3D USCT setup. A center frequency of 2,5 MHz with 1 Mhz bandwidth was chosen. The arrival time of the ultrasound pulse (see equation 3) was calculated via a threshold on the amplitude. The reconstructed image size is 96 $\times$ 96 $\times$ 48 pixel.

The Bézier curve method produces similar results to the fast marching method. The results are represented for one x/y-, y/z- and x/z-slice in Fig. 4. The mean squared error of the Bézier curve technique is 9.98 m/s in comparison to fast marching (11.85 m/s) and straight ray (15.09 m/s). The standard deviation is 3.63 m/s for straight ray, 3.19 m/s for fast marching and 2.88 m/s for Bézier. The maximum deviation reduces from 48.95 m/s for straight ray to 42.20 m/s for fast marching and 41.09 m/s for Bézier respectively. Both bent ray techniques were able to represent the object size more accurately (see difference image presented in Fig. 4), however differ especially in the speed of sound values in the border regions due to blurring by the total variation regularisation required for the sparse aperture. Especially in the upper z regions the boarder is washed out because of the data sparsity.

3.2.2 KIT 3D USCT phantom data

The methods are also applied on a dataset recorded with KIT 3D USCT [14] with 157 transducer array systems (TAS, containing 4 emitters and 9 receivers each) arranged in a semi-ellipsoidal aperture. The imaged phantom consists of a mixture of gelatine, glycerine and propanol with an approximate sound speed of 1560 - 1570 m/s and an olive placed centrally in the phantom. 10 different aperture positions had been recorded resulting in approximately 500 000 A-scans used for reconstruction. The arrival time was calculated via a cross correlation between measurement with object and empty measurement. The reconstructed image size is 96 $\times$ 96 $\times$ 55 pixel for a 25.92 cm $\times$ 25.91 cm $\times$ 14.97 cm image.

The straight and bent ray reconstructions are shown in Fig. 5 for three slices. The object size of the phantom in the Bézier bent ray reconstructed image changed similarly to the fast marching reconstructed image. Two region of interests (ROI) have been selected to evaluate the sound speed values of the gelatine mixture. The median of the Bézier curve reconstructed image in the first ROI is 1580.4 m/s in the second one 1564.9 m/s, for fast marching 1574.5 m/s and 1560.4 m/s in comparison to straight ray with 1560.7 m/s and 1560.4 m/s. In the area of the olive higher sound of speed differences can be observed (see Fig. 5).

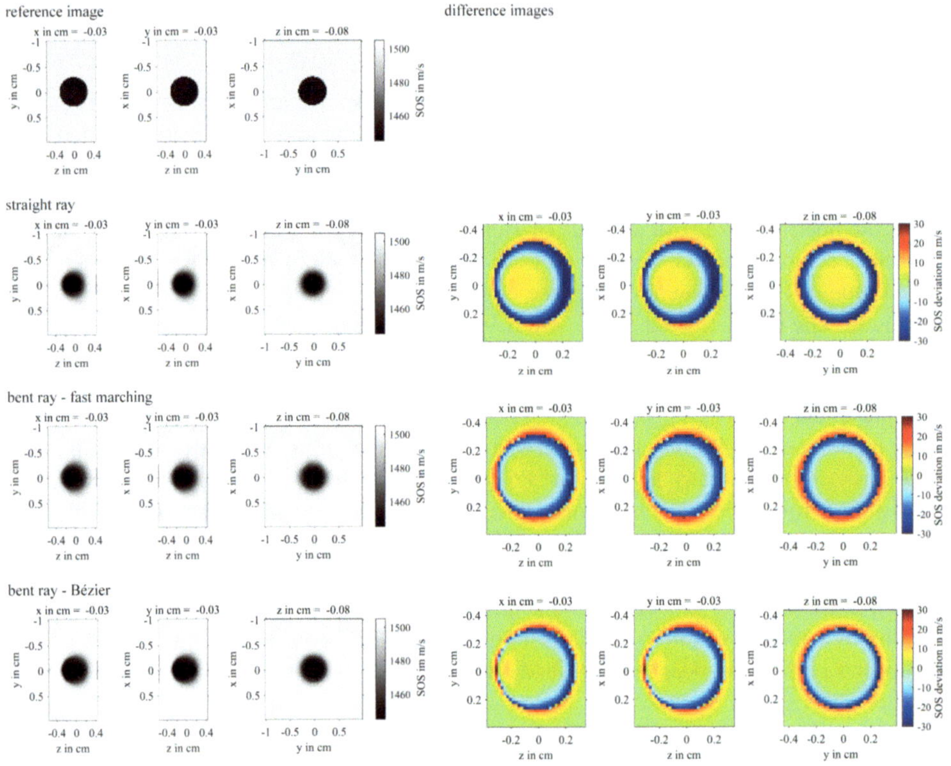

Figure 4: Evaluation of 3D reconstruction with simulated k-wave data. Left column from top to bottom: simulated ground truth, reconstructions with straight ray approximation, fast marching bent ray approximation, and Bézier curve bent ray approximation. The right column shows an enlarged image section of the respective difference to the ground truth with the errors colour coded. Each set of images shows a y/z, x/z and x/y slice from the 3D image.

4 Conclusion

Using Bézier curves for approximating the refraction shows first promising results on 3D simulation data as well as for real data. For simple objects, the Bézier method leads to similar results like the fast marching method. The reconstruction of the object's size improved compared to the straight ray method, the shrinking effect or accordingly the over representation of the object size could be reduced. Bézier curves seem to approximate the main refraction on the object-water interface well. In future, evaluations for more complex objects and real patient data could be carried out to compare the behaviour of the methods. A 3D GPU implementation is currently being developed to compare the computation time for 3D USCT.

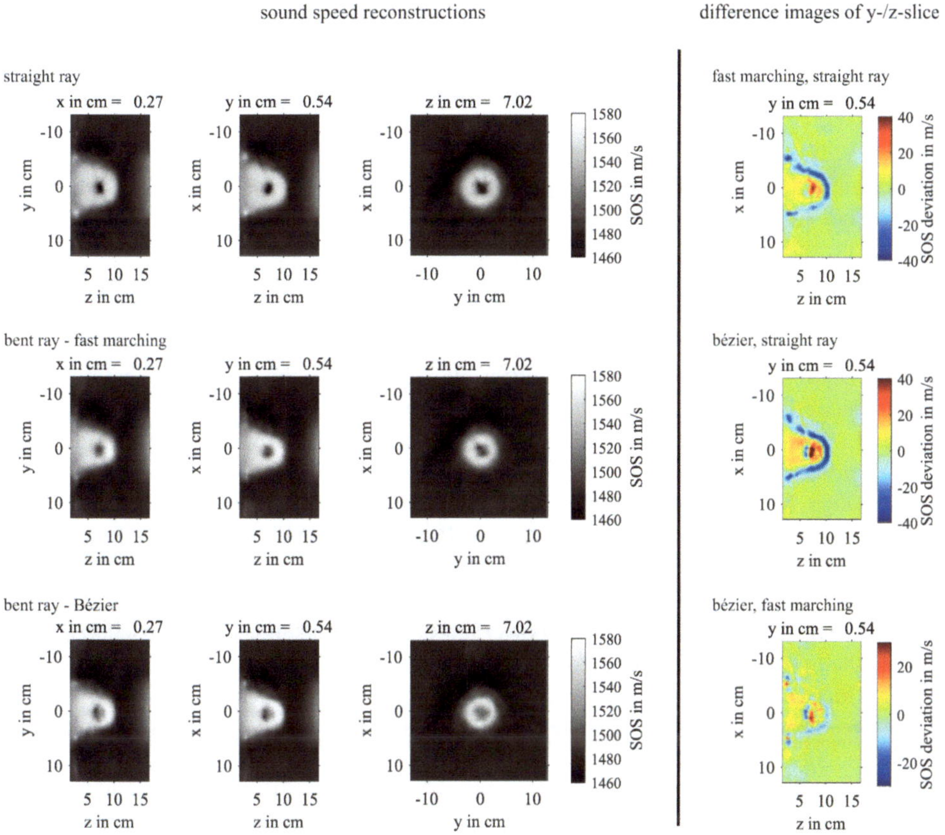

Figure 5: Reconstruction of real experiment data with the KIT 3D USCT. The phantom imaged consists of a mixture of gelatine, glycerine and propanol with an approximate sound speed of 1560 - 1570 m/s. An olive was embedded centrally in the phantom. The images on the left from top to bottom show reconstructed sound speed images with the straight ray approximation, the fast marching bent ray approximation and the Bézier curve bent ray approximation. On the right the sound of speed changes are considered between straight ray and the two bent ray reconstruction.

References

[1] H. Gemmeke, N. V. Ruiter: 3D ultrasound computer tomography for medical imaging. Nuclear Instruments and Methods in Physics Research Section A: Accelerators, Spectrometers, Detectors and Associated Equipment 580(2) (2007), 1057-1065.

[2] J. F. Greenleaf, R. C. Bahn: Clinical imaging with transmissive ultrasonic computerized tomography. IEEE Transactions on Biomedical Engineering, (2) (1981), 177-185.

[3] M. Perez-Liva, J. Udias, J.Camacho, J.Herraiz: Real-Time Ultrasound Transmission Tomography based on Bézier Curves. Proceedings of the International Workshop on Medical Ultrasound Tomography (2018), 89-98.

[4] N. Ruiter: Dreidimensionale Ultraschall-Computertomographie: vom Konzept zur klinischen Anwendung. Habilitation Karlsruher Institut für Technologie, (2016).
[5] J. A. Bærentzen: On the implementation of fast marching methods for 3D lattices (2001).
[6] M. S. Hassouna, A. A. Farag: MultiStencils Fast Marching Methods: A Highly Accurate Solution to the Eikonal Equation on Cartesian Domains. IEEE Transactions on Pattern Analysis and Machine Intelligence 29(9) (2007), 1563–1574.
[7] M. Perez-Liva.: Time domain image reconstruction methods for transmission ultrasound computed tomography. PhD Thesis University Complutense of Madrid (2017).
[8] A. R. Forrest: Interactive interpolation and approximation by bézier polynomials. The Computer Journal, 15(1) (1972), 71–79.
[9] M. Born, E. Wolf: Principles of optics: electromagnetic theory of propagation, interference and diffraction of light. Elsevier (2013).
[10] C. Li: Compressive Sensing for 3D Data Processing Tasks: Applications, Models and Algorithms. PhD thesis, Rice University Houston (2011).
[11] R. Dapp, M. Zapf, N. V. Ruiter: Geometry-independent speed of sound reconstruction for 3D USCT using apriori information. Proceedings IEEE Ultrasonics Symposium (2011), 1403–1406.
[12] B. E. Treeby, J. Jaros, P. Rendell, B. T. Cox: Modeling nonlinear ultrasound propagation in heterogeneous media with power law absorption using a k-space pseudospectral method. The Journal of the Acoustical Society of America 131(6) (2012), 4324–4336.
[13] T. Hopp et al.: Experimental evaluation of straight ray and bent ray phase aberration correction for USCT SAFT imaging. Medical Imaging 2018: Ultrasonic Imaging and Tomography, SPIE (2018) 17.
[14] N. V. Ruiter, et al.: First results of a clinical study with 3D ultrasound computer tomography. IEEE International Ultrasonics Symposium (IUS) (2013), 651-654.

Pseudo-Linear-Frequency-Modulation Pulse Emission and Signal Matching in Ultrasound Computed Tomography System

L. Zhou[1], K. Liu[1], J. Song[1], M. Ding[1,2], and M. Yuchi[1,2]

[1] *Huazhong University of Science and Technology, Wuhan, China*
[2] *WuHan WeSee Medical Imaging Co. Ltd., China*

Abstract

Background

Pseudo-linear-frequency-modulation pulses (pseudochirps) is widely applied as transmitted signals in conventional ultrasound imaging system because of the low cost of the signal generation hardware and better depth of penetration. The spectrum of the pseudoshirps can be modified for better matching to the transducer. Echo signals of pseudochirps after pulse compression may last similar time comparing to the pulse wave but with higher SNR, which has the potential to increase the image quality of the ultrasound computed tomography .

Material and Method

The Lucid system developed in Medical Ultrasound Lab consists of a 2048-elements ring transducer of which the diameter is 220 mm and the center frequency is 3.0 MHz with 70% bandwidth. The pseudochirp wave lasts for 10us and the frequency ranges from 780 kHz to 5.1 MHz. The imaging object is the rectal scan phantom (MODEAL ATS 540) made by CIRS. The phantom has a full 360° acoustic window, and contains five groups of circular targets of varying sizes and depths and five levels of gray scale targets.

Results

The duration of the echo signal of pseudochirp after pulse compression is about 2us, as for the echo signal of two-cycle square wave is about 1.8us. The amplitude of the transmissive waves is higher than the two-cycle square wave. The reconstructed B-mode images show that pseudochirps bring better resolution and contrast.

Conclusion

The results show that the pseudochirp has good potentials in ultrasound computed tomography because of its low cost of hardware, good depth of penetration, high SNR after pulse compression. We will modify the wave and match filter of pseudochirp for better imaging in the future.

A PID Optimizer Approach for Regularizing Quantitative Sound Speed Imaging of an Ultrasound Tomography System

B. Shin[1], G. Ely[1], X. Zhang[1], J. Fincke[1], and B. W. Anthony [1]

[1]*Massachusetts Institute of Technology, Cambridge, USA, E-mail: bh3shin@mit.edu*

Abstract

Although recent quantitative sound speed images generated by full waveform inversion (FWI) provide high resolution images, the overestimation problem hinders convergence and degrades image resolution. In this work, a novel proportional-integral-derivative (PID) optimizer approach updating the sound speed (SS) values is proposed to overcome the convergence issues. By reducing overestimation in FWI, the proposed approach enables to improve the convergence rate and accuracy. The novel gradient descent update rule with the proposed PID approach is validated by comparing to the SS images of limb cross-section in which data set is collected from a custom-made ultrasound tomography scanning system. To evaluate the effect of new convergence parameters, SS reconstructions by the FWI method are separately performed. The results of the momentum effect associated with K_i gain demonstrate that the gradient momentum, history of the previous gradient, significantly contribute to reduce the overestimation of standard deviation values in soft tissue speed of sound. The evaluation results for the damping K_d gain show that the damping parameter can reduce the overshoot phenomena. Overall, the proposed PID approach can reduce both overestimation and overshoot of sound speed images, achieving much robust performance with a faster convergence rate.

Keywords: Full Waveform Inversion, Gradient Descent, PID controller

1 Introduction

Quantitative ultrasound imaging (QUS) illustrating physical property distributions in tissues is a useful medical imaging modality. By mapping those physical properties like strain [1], acoustic attenuation [2], and sound speed [3], the QUS significantly contributes to the detection and management of malignant tissues over the past three decades. Recently, quantitative ultrasound tomography (QUST) presenting sound speed or acoustic attenuation distribution is emerging as a promising QUS imaging tool for breast cancer screening or diagnosis. In the QUST system, full waveform inversion (FWI) technique is a sound speed inversion algorithm that minimizes the data residual between simulated and observed ultrasound pressure fields. Due to the consideration of high-order diffraction and refraction effects, the FWI methods can offer higher spatial resolution images than ray-based imaging and sound speed inversion methods [4, 5] which estimate time-of-flight (TOF) data of each emitter-receiver pair.

FWI techniques using gradient descent methods to update sound speed maps suffer from overestimation, the values become inaccurate and noisy and it contributes slow convergence. Similarly, overestimation and overshoot problems greatly decreasing the learning rates of gradient descent-based optimization methods used in machine learning training. The learning rate in the machine learning training is an important hyper-parameter to quickly train a deep neural network data set [6]. To improve the machine learning rate, many advanced optimization algorithms have been currently introduced and they can be categorized into two groups: (1) hand-tuned learning rate optimizers such as stochastic gradient descent, (SGD) [7] and SGD Momentum [8] and (2) automatically tuned rate optimizer such as AdaGrad [9], RMSProp [10] and Adam [11]. In practice, SGD Momentum and Adam optimizer are the most widely used optimization algorithms due to their robustness, but overshoot is still problematic. The recent findings by An et al. [12] show that the overshooting problem caused by the gradient history hinders the convergence of SGD-Momentum and increases training time. Also, the adaptation of the classic proportional-integral-derivative (PID) controller for optimizing learning rates achieves fast convergence rate and reduces the over-shoot phenomena [12].

Inspired by the idea from An et al. [12], a robust gradient descent update rule from the PID controller approach is investigated to improve the convergence rate of the FWI technique and accuracy of the final sound speed map. This paper is organized as follows. Section 2 briefly reviews the conventional FWI technique and introduces the proposed PID approach. In Section 3, we describe an acoustic data acquisition method using an in-house single element 3D ultrasound tomography system [13] (Fig. 1(a)) and practical FWI implementation are briefly introduced for real data. In Section 4, we invert sound speed images using conventional FWI and the proposed PID method.

2 Full waveform inversion and PID approach

In this section, theorical background of full waveform inversion is briefly introduced and we discuss the relationship between the current gradient descent-based update equation in FWI and the industrial PID controller. From this relationship, we propose the novel SS update equation using the PID optimizer.

2.1 Full waveform inversion technique

Here we briefly describe the conventional FWI to reconstruct the spatial distribution of sound speeds. In this study, the observed time series signals are collected from a custom-made ultrasound tomography scanning system described in Section 3. By minimizing the mean square error between the observed times series and the simulated time series produced by simulating the wave equation, the FWI method enables us to reconstruct the sound speed map. In the conventional FWI, the objective function is typically given by:

$$E(c,\rho) = \frac{1}{2}\sum_{n=1}^{N} \|F_n(c,\rho,s_n) - d_n\|^2, \tag{1}$$

where E is the mean square error between the simulated and the observed data, F_n is the wave equation operator which outputs a simulated time series of speed sound with units of Pa, s_n is the source for the n-th transmission and d_n is the observed time series data from the n-th transmission with units of Pa and c_o is the initial sound speed distribution. An acoustic simulation code, k-wave toolbox [14, 15], is used to solve the wave equation operator. The acoustic simulation toolbox includes the effects of sound speed, c, in units of m/s, density, ρ, in units of kg/m^3. The FWI reconstruction is similar to the unconstrained nonlinear inversion problem that is given by,

$$\min_{c,\rho} E(c,\rho) = \min_{c,\rho} \frac{1}{2}\sum_{n=1}^{N} \|F_n(c,\rho,s_n) - d_n\|^2. \tag{2}$$

In the conventional FWI, local optimization algorithms are generally used to solve eq. (2). According to the adjoint-state method, the gradient of objective function can be given by [16, 17, 18],

$$g_i(c|F_n,d_n) = \frac{\partial E(c,\rho)}{\partial c} = \frac{2}{c^3}\sum_{n=1}^{N}\sum_{t=0}^{T} \frac{\partial^2 p_n}{\partial t^2} p_n^*(t-T), \tag{3}$$

where p_n and p_n^* represent full wavefield and back-propagated residual wavefield, respectively. The back-propagated residual wavefield, p_n^*, is calculated by setting the residual data as virtual sources and performing forward modeling from the last time point to first time point. After obtaining the gradient in eq. (3), speed sound is updated iteratively by the following update equation:

$$c_{k+1} = c_k + \gamma d_k, \tag{4}$$

where γ is the step length and d_k is update direction when employing the steepest descent method, $d_k = -g_k$, where $g_k = \partial E/\partial c$. Newton type of algorithms (i.e., BFGS, L-BFGS) [19] are also employed to find the update direction, in which case is $d_k = -H_k^{-1} g_k$, where H_k is the Hessian matrix.

2.2 PID Optimizer Approach

In this section we briefly derive the PID like update for full waveform inversion. We first give a brief overview of the conventional PID controller and then adapt the update to FWI.

2.2.1 PID controller

Although many advanced control algorithms have previously been proposed, PID controllers are still widely used to control feedback systems in many industry fields because they are simple, robust and easy to use [20]. A PID controller determines its correction $u(k)$ to the system based on the proportional (P), integral (I), and derivative (D) terms of the error $e(k)$ between the desired optimal output and a measured system output and the correction term is mathematically given by,

$$u(k) = K_p e(k) + K_i \int_0^k e(k)\,dt + K_d \frac{d}{/} dke(k), \tag{5}$$

where K_p, K_i, and K_d are the proportional (P), integral (I), and derivative (D) gain coefficients determining the contributions of present, past and future errors to the current correction.

2.3 Gradient Descent update as a P controller

As described in eq. (4), the steepest gradient descent method is a typical update equation minimizing the differences between the simulated and observed model. By considering the gradient term (g_k) as error $e(k)$ in eq. (5), the gradient descent only uses the present gradient to update the weights. Therefore, the typical gradient descent update equation in FWI is a type of P controller with the proportion gain γ. Thus, the update update can be written as,

$$c_{k+1} = c_k - \gamma e(k) = c_k - K_p e(k). \tag{6}$$

2.4 Momentum Approach (PI Controller)

Similarly, the gradient descent (GD) based algorithm is also representative of the optimization algorithms used in deep neural networks [6]. In neural network, stochastic gradient descent (SGD) with momentum [8] increases training rates compared with SGD [7], and its parameter update is given by [12],

$$\begin{aligned} V_{k+1} &= \alpha V_k - \gamma g_k, \\ c_{k+1} &= c_k + V_{k+1}, \end{aligned} \tag{7}$$

where, V_k is the accumulation of the history gradient, and $\alpha \in (0,1)$ is the rate of moving average decay. The introduction of decay term α decreases the impact of far away models on the gradient and reduces noise. In this framework, the GD with momentum term can be considered a PI controller.

2.4.1 Full PID controller Approach

The momentum term in the update equation accumulates the history of gradients. The accumulation of the previous gradient terms changes the descending direction to more quickly minimize the objective function. The history gradients lag the update of weights and such a phenomenon caused by the previous gradient accumulation is called overshoot. To build a simple PID optimizer, a derivative term is added to GD-momentum and the PID optimizer updates are given by [12],

$$\begin{aligned} V_{k+1} &= \alpha V_k - \gamma g_k, \\ D_{k+1} &= \alpha D_k + (1-\alpha)(g_k - g_{k-1}), \\ c_{k+1} &= c_k + V_{k+1} + K_d D_{k+1}, \end{aligned} \tag{8}$$

where D_k is the derivative (change of gradient) at the iteration k. The proposed PID optimizer uses more "future" error term and reduces the overshoot problem.

3 Method

Experimental data acquisition system using a custom-made ultrasound tomography system and the implementations of full waveform inversion we use are introduced in the following section.

3.1 Scanning system and experimental data Acquisition

We used a custom-made single element ultrasound tomography system [13] to acquire the acoustic data sets from a beef shank. As shown in Fig. 1 (a), two submerged single element ultrasound transducers move independently and radially over the entire 360° aperture. The transducers can also translate approximately 7 cm vertically. The tranducers are assembled on top of a ring bearing to scan a submerged object in a water tank. The designed system can scan submerged objects in either pulse-echo or through transmission modes. Further information regarding calibration procedure, mechanical system design, data acquisition system is described in [13].

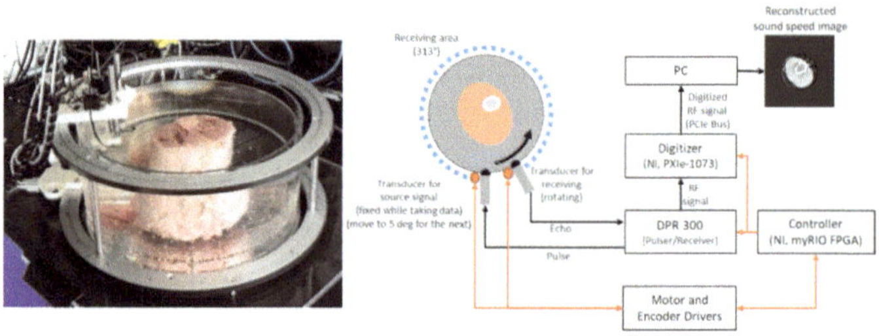

(a) (b)

Figure 1: Ultrasound tomography scanning system: (a) ultrasound tomography system with a water tank, (b) schematic of ultrasound tomography scanning system.

A low frequency pulser/receiver (DPR300, Imaginant, Pittsford, NY, USA) is used to drive the transmitting transducer and receive the analog signal from the receiving transducer (Fig. 1 (b)). The analog signal receiving from DPR 300 is digitized at 50 MHz (NI PXIe-1073, National Instruments, Austin, TX, USA) and streamed to a PC where the signal is saved for processing. Dual element (bi-static) scans consisting of 72 transmit locations equally spaced 5° apart resulting in a 360° synthetic aperture are employed to collect the experimental data sets. At each transmit location, the radio frequency (RF) signals receiving by another single US transducer are sequentially acquired at 1333 equally distributed spatial locations (0.2345° apart) resulting in a 313° receive aperture (Fig. 1 (a)). In this experiment, 1.0 MHz broad band custom made transducers are employed. The source (transmit) transducer is the same design as those used for the bi-static scans. The receiving transducer is 1.47 mm diameter disk shaped receiver which can be considered a point receiver at 1.0 MHz. The source transducers were measured with a calibrated a hydrophone to ensure the scans do not exceed the food and drug administration (FDA) standards for ultrasound use on humans [13]. All scanning procedures described were approved by the MIT committee on animal care.

3.2 Implementations of full waveform inversion

A MATLAB-based FWI toolbox based on the conventional FWI (C-FWI) and the FWI with the proposed PID update equation is implemented as briefly described in Fig. 2. The C-FWI algorithm computes the gradient of the objective function using adjoint-state method in eq. (3) and uses steepest gradient descent method to update the sound speed distributions at each iteration. During the computation of gradient values of the objective function, the C-FWI algorithm requires calculating the wave equation two times by number of source data sets (transmit data sets) for each iteration. In this work, the k-wave toolbox [14, 15] is used to compute the forward wave equation. In practice, the C-FWI algorithm is solved with multiple cores using parallel computing toolbox in the MATLAB. We compute the average gradient values of 72 independent gradient fields from the different transmits, the average gradient values are divided by the maximum peak gradient value at the first iteration. At each iteration, a line search algorithm determines step size as the proportional gain γ (K_p). Lastly, the sound speed distributions of each iteration is updated using the update equations of either C-FWI or the C-FWI with the proposed PID controller approach considering history gradient and future prediction of gradient terms (K_i and K_d) as described in eq. (4) or eq. (8).

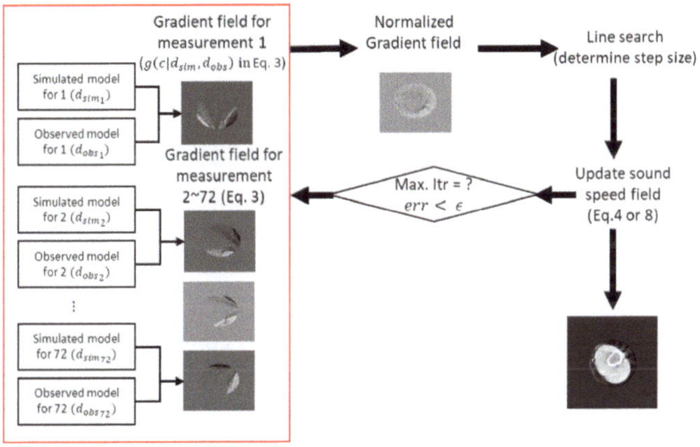

Figure 2: Flow chart of the full waveform inversion algorithm chart.

4 Results

Sound speed values of the soft tissue in the beef shank are compared to evaluate the proposed PID optimizer approach. Before evaluating the proposed full PID optimizer approach, the momentum approach (PI approach) which considers history of the previous gradient are first evaluated by comparing average and standard deviation values of the sound speed maps. In the same manner, the proposed PID optimizer approach is also assessed to understand the impact of the damping term on the update equation.

4.1 Evaluation of PI Approach with Momentum Change (K_i gain)

The momentum parameter α in eq. (7) as the K_i gain in PI optimizer approach is tuned from 0 to 0.3 to determine the optimal momentum value. As shown in Fig. 3 (a), the average values after 7th iteration are converged to near 1555 m/s which agrees with the ground truth value we previously measured (1556 m/s) [12]. The average sound speed values from all the PI optimizer approaches (K_i 0.1 - 0.3) increase more rapidly than those from conventional FWI (zero K_i gain). When we increase the K_i gain, the average sound speeds converge quickly within 6 iterations at a K_i of 0.3. However, the standard deviation of sound speeds increases after the 7th iteration, making the sound speed values oscillate around the ground truth value. These errors are amplified with subsequent iterations. The standard deviation plots (Fig. 3 (b)) show that the larger momentum, K_i, gain contributes to the stabilization of standard deviation values after 6th iteration.

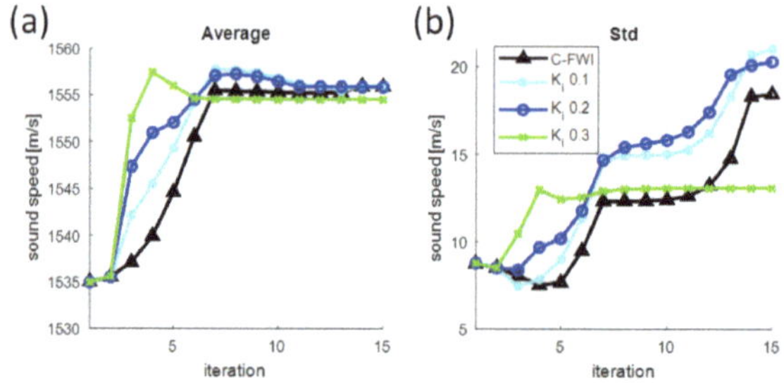

Figure 3: Average and standard deviation plots: (a) average values, and (b) standard deviation values of sound speed in soft tissue with K_i gain change.

Sound speed distribution maps are compared in Fig. 4 (a) to (c). When the K_i gain increases, the sound speed distribution maps become smoother and contain less high frequency noise. However, the image contrast from K_i 0.3 (Fig. 4 (c)) decreases compared to inversions with lower K_i values. Sound speed values of the horizontal line at the 15th iterations are shown in Fig. 4(d). The larger K_i values decrease the standard deviation in soft tissue and reduce the over-estimation between the boundaries between soft tissue, bone, and water.

4.2 Evaluation of PI Approach with Momentum Change (K_i gain)

The role of damping gain (K_d) in a typical PID controller is to decrease overshoot effect. To investigate the damping effect of the proposed PID optimizer approach in FWI, average and standard deviation of sound speed values are plotted in Figure 5. In this section, the K_i gain is fixed to 0.25 and the damping coefficient (K_d in eq. (8)) are set to three levels as the none (0), moderate (0.5), and big damping gain (4), respectively. The average sound speed values from none and moderate damping cases converged to 1555 m/s. However, values from big damping gain reach a slightly higher value (1557 m/s), and the overshoot effect

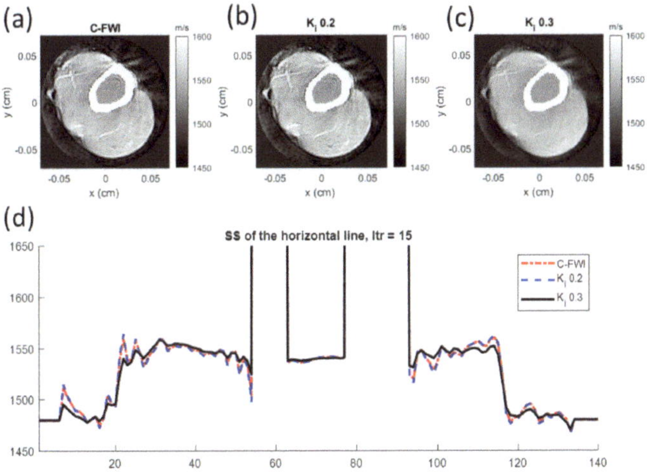

Figure 4: Sound speed distribution maps and plots at the 15th iteration: (a) Sound speed (SS) map generated by conventional full waveform inversion (C-FWI), (b) SS map produced by applying K_i gain 0.2, and (c) SS map from K_i gain 0.3, (d) sound speed plots of horizontal line.

is significantly decreased for the biggest damping gain (Fig. 5 (a)). As for the standard deviation values of sound speed, both moderate and big damp converge to 12.5, 13 m/s, respectively. The standard deviation values for the none damping is continuously increasing with the iteration and is unstable (Fig. 5 (b)).

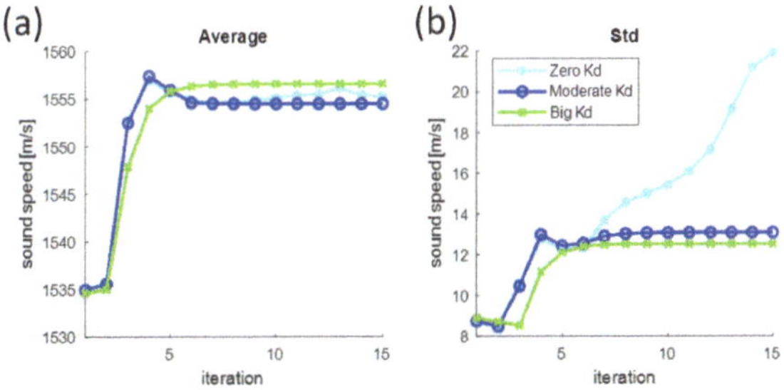

Figure 5: Average and standard deviation plots: (a) average of sound speed, (b) standard deviation of sound speed in soft tissue with K_d gain change.

Fig. 6 (a) to (c) show the sound speed distribution maps for several damping parameters K_d. Like the effect of K_i gain, the sound speed distribution maps become smoother and show less noise as the damping increase. However, the image contrast from big damping (Fig. 6 (c)) is significantly degraded. The sound speed plots of the horizontal line with the moderate K_d depicts the smooth results, but the cross section from the big K_d inversion shows the over-smoothing trend (Fig. 6 (d)).

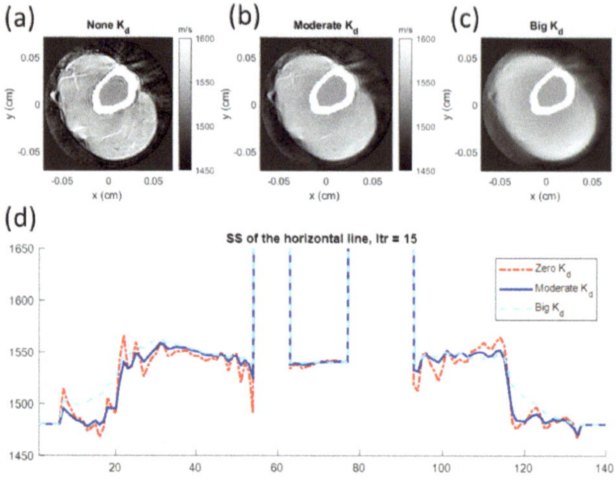

Figure 6: Sound speed distribution maps and sound speed plot: (a), (b), and (c) Sound speed map generated by the proposed PID approach with zero K_d gain, moderate and big K_d gain, respectively, (d) sound speed plot of horizonal line.

5 Conclusions

A novel robust sound speed update algorithm for FWI is proposed to reduce overestimation and to produce accurate quantitative images. Here we have implemented and investigated the novel update algorithm with a PID optimizer approach that utilizes the present, past, and change information of gradients. The evaluation results for the PI controller approach additionally considering K_i gain significantly improves the overestimation problem of standard deviation values of the sound speed map and increases converges to the steady-state values. In the same manner, the PID optimizer approach using K_p, K_i and K_d gain reduces overshoot phenomena and stabilizes the standard derivation values of the sound speed in the soft tissue of beef shank. Therefore, the convergence rate and stability of the new sound speed update algorithm are greatly improved and the proposed PID approach can produce much smoother sound speed images compared to conventional FWI. In future work, we will investigate other auto-learning rate algorithms in machine learning optimization (i.e., AdaGrad [9], Adam optimization [11]) to further accelerate the convergence speed of the FWI technique without using the line search algorithm that adaptively determines step size of the update equation. Moreover, this novel FWI toolbox with the proposed PID approach will be extended to our new custom-made ultrasound tomography scanning system using multi-arms robots.

Acknowledgements

Bonghun Shin has been supported by NSERC-PDF awarded from the Natural Sciences and Engineering Research Council of Canada (NSERC) throughout the duration of this work.

References

[1] B. Shin, S. Jeon, J. Ryu, and H. J. Kwon, "Compressed sensing for elastography in portable ultrasound," *Ultrasonic Imaging*, vol. 39, no. 6, pp. 393–413, 2017. PMID: 28670990.

[2] N. V. Ruiter, M. Zapf, T. Hopp, H. Gemmeke, K. W. A. van Dongen, J. Camacho, J. L. Herraiz, M. P. Liva, and J. M. Udías, "USCT reference data base: conclusions from the first SPIE USCT data challenge and future directions," in *Medical Imaging 2018: Ultrasonic Imaging and Tomography* (N. Duric and B. C. Byram, eds.), vol. 10580, pp. 170 – 176, International Society for Optics and Photonics, SPIE, 2018.

[3] K. Wang, T. Matthews, F. Anis, C. Li, N. Duric, and M. A. Anastasio, "Waveform inversion with source encoding for breast sound speed reconstruction in ultrasound computed tomography," *IEEE Transactions on Ultrasonics, Ferroelectrics, and Frequency Control*, vol. 62, no. 3, pp. 475–493, 2015.

[4] R. G. Pratt, L. Huang, N. Duric, and P. Littrup, "Sound-speed and attenuation imaging of breast tissue using waveform tomography of transmission ultrasound data," in *Medical Imaging 2007: Physics of Medical Imaging* (J. Hsieh and M. J. Flynn, eds.), vol. 6510, pp. 1523 – 1534, International Society for Optics and Photonics, SPIE, 2007.

[5] O. Roy, I. Jovanović, A. Hormati, R. Parhizkar, and M. Vetterli, "Sound speed estimation using wave-based ultrasound tomography: theory and GPU implementation," in *Medical Imaging 2010: Ultrasonic Imaging, Tomography, and Therapy* (J. D'hooge and S. A. McAleavey, eds.), vol. 7629, pp. 177 – 188, International Society for Optics and Photonics, SPIE, 2010.

[6] I. Goodfellow, Y. Bengio, A. Courville, and Y. Bengio, *Deep learning*, vol. 1. MIT press Cambridge, 2016.

[7] L. Bottou, "On-line learning and stochastic approximations," in *In On-line Learning in Neural Networks*, pp. 9–42, Cambridge University Press, 1998.

[8] I. Sutskever, J. Martens, G. Dahl, and G. Hinton, "On the importance of initialization and momentum in deep learning," in *Proceedings of the 30th International Conference on Machine Learning* (S. Dasgupta and D. McAllester, eds.), vol. 28 of *Proceedings of Machine Learning Research*, (Atlanta, Georgia, USA), pp. 1139–1147, PMLR, 17–19 Jun 2013.

[9] J. Duchi, E. Hazan, and Y. Singer, "Adaptive subgradient methods for online learning and stochastic optimization," *J. Mach. Learn. Res.*, vol. 12, pp. 2121–2159, July 2011.

[10] K. S. G. Hinton, N. Srivastava, "Lecture 6a overview of mini-batch gradient descent," 2012. Coursera Lecture slides https://class.coursera.org/neuralnets-2012-001/lecture.

[11] D. P. Kingma and J. Ba, "Adam: A method for stochastic optimization," in *3rd International Conference on Learning Representations, ICLR 2015, San Diego, CA, USA, May 7-9, 2015, Conference Track Proceedings* (Y. Bengio and Y. LeCun, eds.), 2015. arXiv: 1412.6980.

[12] W. An, H. Wang, Q. Sun, J. Xu, Q. Dai, and L. Zhang, "A PID Controller Approach for Stochastic Optimization of Deep Networks," in *2018 IEEE/CVF Conference on Computer Vision and Pattern Recognition*, pp. 8522–8531, 2018.

[13] X. Zhang, J. Fincke, A. Kuzmin, V. Lempitsky, and B. Anthony, "A single element 3D ultrasound tomography system," in *2015 37th Annual International Conference of the IEEE Engineering in Medicine and Biology Society (EMBC)*, pp. 5541–5544, 2015.

[14] B. E. Treeby and B. T. Cox, "k-Wave: MATLAB toolbox for the simulation and reconstruction of photoacoustic wave fields," *Journal of Biomedical Optics*, vol. 15, no. 2, pp. 1 – 12, 2010.

[15] B. E. Treeby, J. Jaros, A. P. Rendell, and B. T. Cox, "Modeling nonlinear ultrasound propagation in heterogeneous media with power law absorption using a k-space pseudospectral method," *The Journal of the Acoustical Society of America*, vol. 131, no. 6, pp. 4324–4336, 2012.

[16] C. Zhou, W. Cai, Y. Luo, G. T. Schuster, and S. Hassanzadeh, "Acoustic wave-equation traveltime and waveform inversion of crosshole seismic data," *Geophysics*, vol. 60, no. 3, pp. 765–773, 1995.

[17] K. Wang, T. Matthews, F. Anis, C. Li, N. Duric, and M. A. Anastasio, "Waveform inversion with source encoding for breast sound speed reconstruction in ultrasound computed tomography," *IEEE Transactions on Ultrasonics, Ferroelectrics, and Frequency Control*, vol. 62, no. 3, pp. 475–493, 2015.

[18] M. Pérez-Liva, J. L. Herraiz, J. M. Udías, E. Miller, B. T. Cox, and B. E. Treeby, "Time domain reconstruction of sound speed and attenuation in ultrasound computed tomography using full wave inversion," *The Journal of the Acoustical Society of America*, vol. 141, no. 3, pp. 1595–1604, 2017.

[19] D. C. Liu and J. Nocedal, "On the limited memory bfgs method for large scale optimization," *Mathematical programming*, vol. 45, no. 1, pp. 503–528, 1989.

[20] L. Wang, "New frequency-domain design method for PID controllers," *IEE Proceedings - Control Theory and Applications*, vol. 142, pp. 265–271(6), July 1995.

Transceiver ASIC in HVCMOS Technology for 3D Ultrasound Computer Tomography

R. Blanco[1], R. Leys[1], K. Schlote-Holubek[1], L. Becker[1], M. Zapf[1], P. Steck[1], H. Gemmeke[1], N. V. Ruiter[1], and I. Perić[1]

[1] *Karlsruhe Institute of Technology (KIT), Karlsruhe, Germany, E-mail: roberto.blanco@kit.edu*

Abstract

3D Ultrasound Computer Tomography (3D USCT) is an imaging method for the early detection of breast cancer. It provides three-dimensional multimodal images of the breast. The new 3D USCT device developed currently at Karlsruhe Institute of Technology contains more than two thousand ultrasound transducers placed in a water-filled aperture where the patient submerges one breast. The ultrasound transducers are grouped as transducer array systems (TAS) of 18 receiver (RX) and transmitter (TX) elements. The transducer front-end electronics contain high-voltage (HV) and low-voltage (LV) amplifiers and switches which are implemented as an application-specific integrated circuit (ASIC). This contribution presents a patented mixed signal, multichannel, transceiver ASIC developed in a commercial 350 nm high-voltage CMOS (HV-CMOS) process. The HV-CMOS process provides low-voltage and high-voltage transistors that can be combined on the same substrate. The HV transistors can sustain voltages up to 120 V.

Keywords: application-specific integrated circuit, front-end electronics, medical imaging

1 Introduction

The previous prototype, 3D USCT II is displayed in Figure 1. For our new prototype, 3D USCT III, a larger opening angle and an enlarged aperture with a diameter of 36 cm will increase [3] the region of optimal 3D point spread function [1]. The aperture of ultrasound transducers is hemispherical. The hemispherical aperture can be rotated and translated (for USCT II and III) to increase the number of virtual transducer positions. The measured signals are the so-called A-scans (amplitude scans). Approximately spherical waves in emission and approximately spherical reception are used for full 3D imaging. The center frequency of the ultrasound pulses are approximately 1 to 2.5 MHz. In this work we will present an

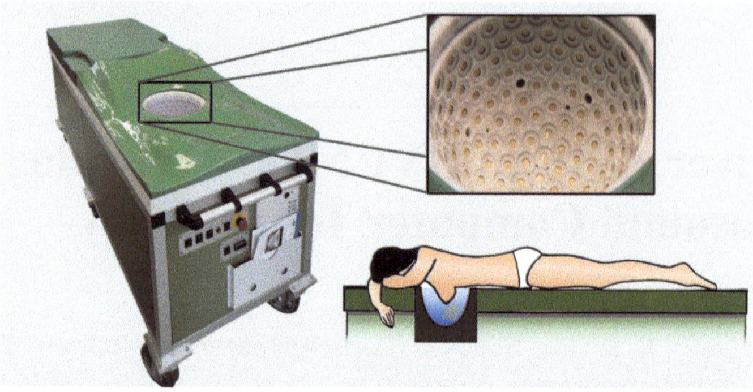

Figure 1: USCT II patient bed (left), transducer aperture (top right) and patient position (bottom right).

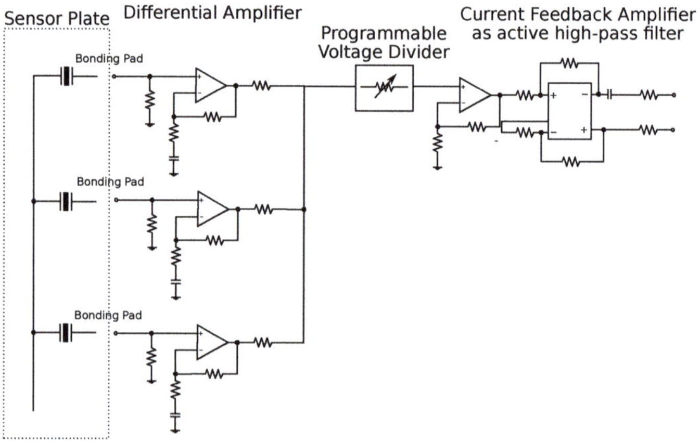

Figure 2: USCT II TAS receiver. Sensor plate, differential amplifier, voltage divider and current feedback amplifier as active high-pass filter [2].

integrated front-end solution. Using the transducer as emitter (TX) and receiver (RX). The TX applies high-voltage to the piezoelectric transducer generating ultrasound pulses. In the other direction, the RX amplifies the low signal amplitudes. An ASIC will be presented to solve these tasks. The development was started with the first generation ASIC consisting of one high-voltage (TX) and low-voltage channel (RX). The channel replaces the discrete step up converter, optocouplers and amplifiers of TAS USCT II emitter and receiver (see Figures 2 and 3). The ASIC features nine channels as required for USCT III and using a 3:1 multiplexer. The ASIC can be programmed with an micro-controller using the serial peripheral interface (SPI). A digital SPI decoder was also implemented in the ASIC.

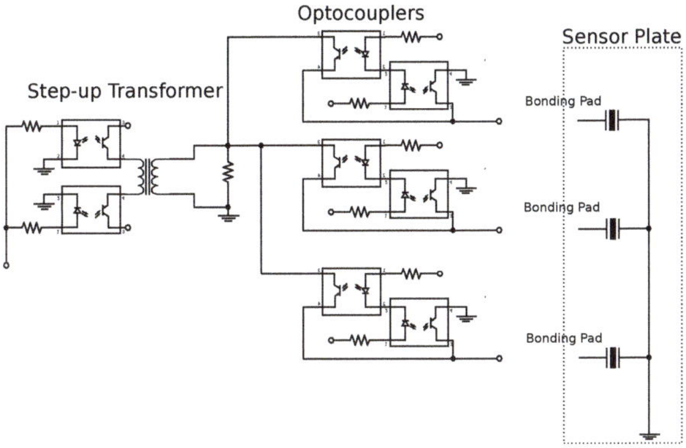

Figure 3: USCT II TAS emitter. Sensor plate, step up transformer, optocouplers [2].

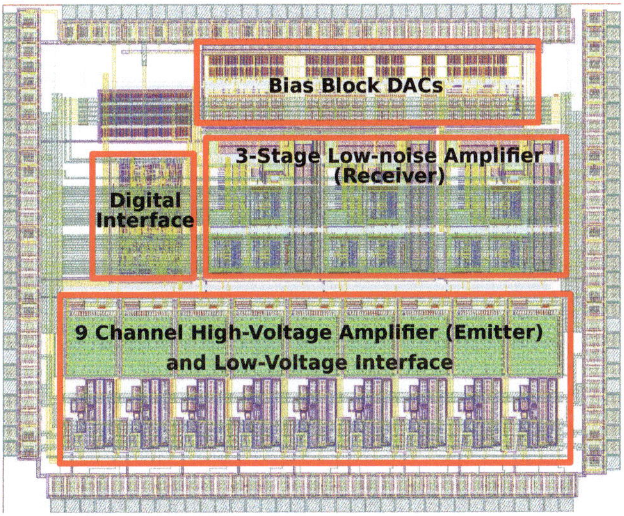

Figure 4: The ASIC consists of 3-stage low-noise amplifiers, noine channel high-voltage amplifier and digital interface and bias block.

2 ASIC Architecture

The ASIC (see Figure 4) consists of nine high-voltage amplifiers as TX elements and three low noise low-voltage amplifiers as RX elements. The high voltage amplifier is designed as an inverting amplifier with a gain of 20. It can generate signals with an amplitude up to 120 V. The 3-stage low noise amplifier is designed for amplification of signals from 10 μV to 10 mV with variable gain from 100 to 10000. In addition, the ASIC contains a digital interface for its configuration. This digital block implements the SPI decoder. The RX signals of 100 μV were amplified by the LV amplifiers up to 300 mV with a gain factor of around 3000.

Technology	AMS 350 nm HV-CMOS
Chip size	3.3 mm x 4.1 mm
High-voltage amplifier	Class AB amplifier (120 V transistors)
Low-noise amplifier	Inverting voltage amplifier, CR-RC shaper (3-stage)
Power consumption (high-voltage channel)	10 mW
Bias current	10 μA
Signal rise time	120 ns
3dB bandwidth	300 kHz - 6 MHz
Gain (high-voltage amplifier)	20
Gain (low-noise amplifier first stage)	40
Output related noise	121 μV
Input related noise	4 μV

Table 1: ASIC main characteristics [4].

The use of the ASIC offers the following advantages for USCT over a discrete component solution:

1 Low power design to avoid heating of the transducers

2 Small size design

3 Crosstalk reduction

4 More RX/TX channels

5 Cost effective solution

Table 1 shows the main characteristics of the ASIC. A simplified overview of the one input channel is depicted in Figure 5. One channel has an emitter stage, tune capacitances, 3-stage low-noise amplifier and switches to enable TX or RX. Figure 6 displays the simplified multichannel architecture with nine input channels and three output channels. The output channels are multiplexed with a 3:1 multiplexer stage.

2.1 High-Voltage Amplifier (TX)

To implement the emitter stage the step-up transformer was replaced by a transconductor and class AB amplifier stage. The transconductor performs a voltage-to-current conversion. The transconductor consists of a cascode differential amplifier (Figure 7) with two input transistors, cascode transistors and current mirror. The difference in their current values is proportional to the difference in the input voltages. Figure 7 shows a simplified schematic of the real implementation. Including the transconductor offset generation. The high-voltage

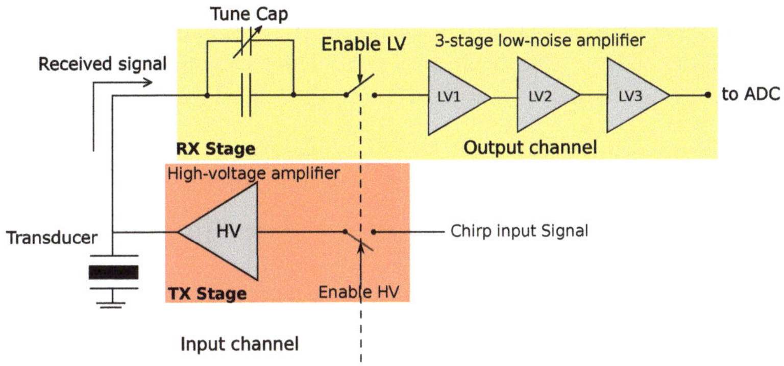

Figure 5: Simplified block schematic of one channel composed of an TX, RX stage, tune capacitances, HV and LV enable switches.

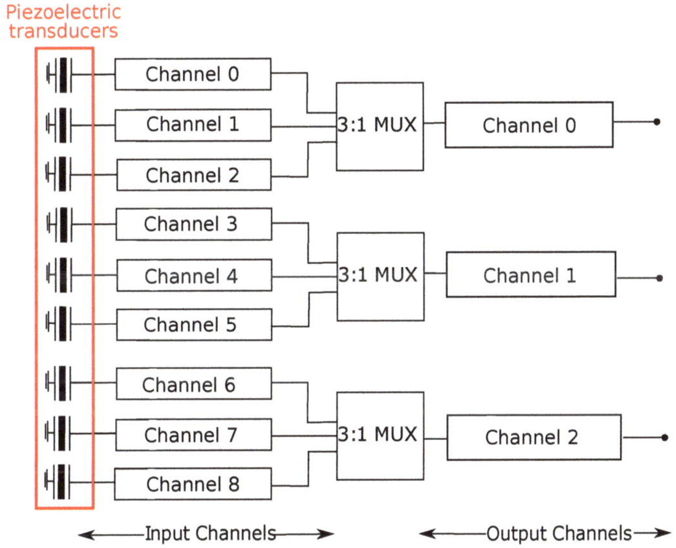

Figure 6: Simplified block diagram of the multichannel architecture [4].

amplifier requires an offset as voltage reference. In a next step the transconductor was connected to a source follower stage. To improve the circuit the current source must be replaced by a transistor. This type can be described as class B output stage. A class B stage has a distortion around the input voltage $V_{in} = 0$. In order to avoid the distortion an improved source follower stage with two diode-connected transistors was introduced. A stable voltage is achieved which shifts the operating points and the class B becomes a low distortion class AB amplifier (see Figure 8) [4]. The class A part of the amplifier has high gain and high linearity. This kind of amplifier has a 2π conduction angle. That means the amplifier remains active for 360-degree signal and the complete input signal is used. The class B amplifier consists of two MOSFET transistors (NMOS and PMOS). Each transistor is biased during the positive and negative of a sinusoidal input signal. Here the signal gets pushed or pulled. As mentioned before, to overcome the distortion problem the combination of both structures are used [4].

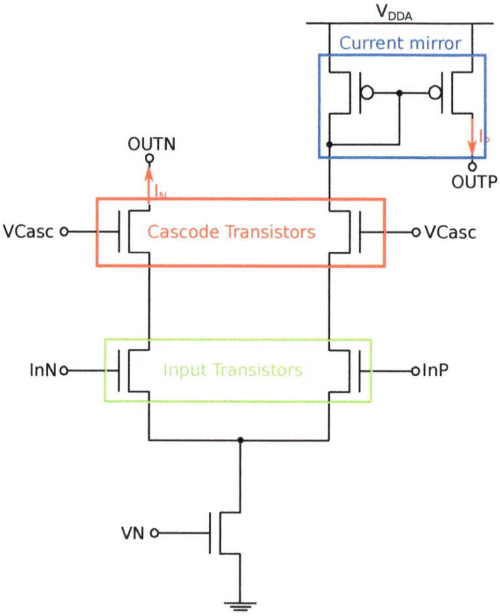

Figure 7: Transconductor circuit based on a cascode differential amplfier circuit (simplified) [4].

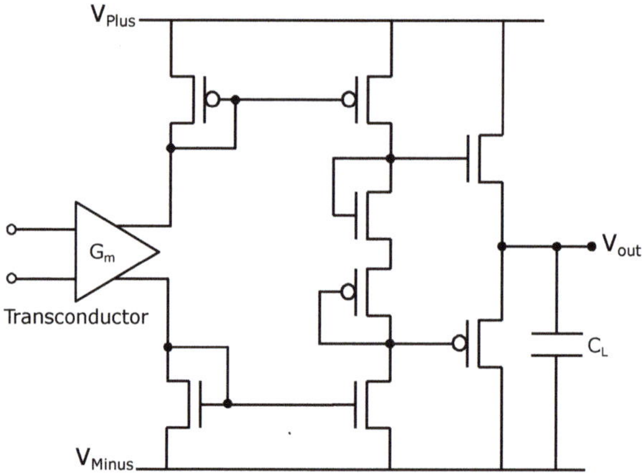

Figure 8: The high-voltage consists a class A and B stage in combination [4].

To verify the functionality of the emitter, various simulations like transient, large signal analysis, small signal analysis and noise were performed. The excitation signal used in the simulations is the linear chirp signal. Such signal input is set to receive analog high-voltage signals, which can be provided by an external high-voltage transmitter. The chirp frequency increases linearly with time, so the frequency response of this signal has a constant amplitude. In the simulation a high-voltage of ±48 V was used. The amplifier with a closed loop gain of 20 can deliver an output of 96 V peak-to-peak. The closed loop was realized as DC feedback. The TX emitter is implemented as an inverting amplifier (see Figure 9) with three outputs. Two of them are fed to the feedback with a feedback capacitance to increase the stability. To simulate the piezoelectric transducer an equivalent circuit was used as load for the TX emitter. An impedance analyzer was used to determine the LCR parameter. It contains an LCR series network and three parallel capacitances forming the capacitive load. This type of circuit can be handled as series resonant circuit. Note, to guarantee maximum gain and high linearity a parameter sweep of all bias parameters must be performed. Figure 9 shows the simulation circuit of the TX emitter as inverting amplifier topology [4].

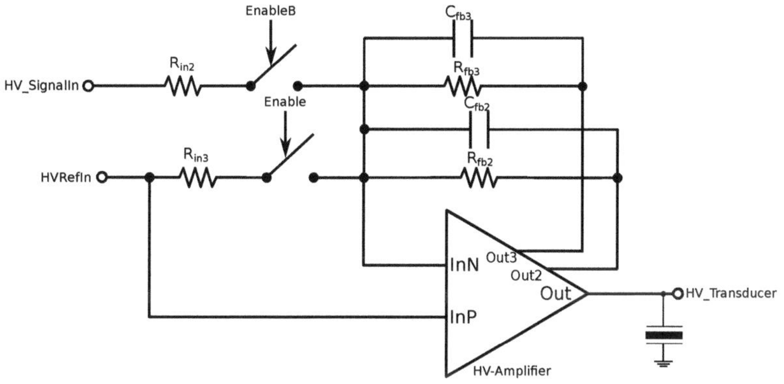

Figure 9: TX emitter amplifier works as an inverting amplifier. The feedback capacitance are used to improve the stability.

2.2 Low-Voltage Amplifier (RX)

The ultrasonic signal received from the piezoelectric transducer has low signals in order of 1 μV to 100 μV. To amplify such small signals a 3-stage low-noise amplifier was designed (see Figure 10). A 3:1 multiplexer selects one of three input receivers. The first stage achieves a gain of 100 where a inverting voltage amplifier with capacitive loads was chosen to achieve low-noise. The second and third stages are composed of CR-RC shaper topology. The simulated gain in the second stage can be chosen between 1-16 and for the third stage between 2-64. The amplifier is stable up to an amplification value of 5000. To set the gain, the tune capacitance can be switched in parallel to increase the gain factor. In the third stage the input capacitance can also be tuned. The outputs from stages 2 and 3 are fed into the a output selection logic. This logic is realized as a multiplexer [4].

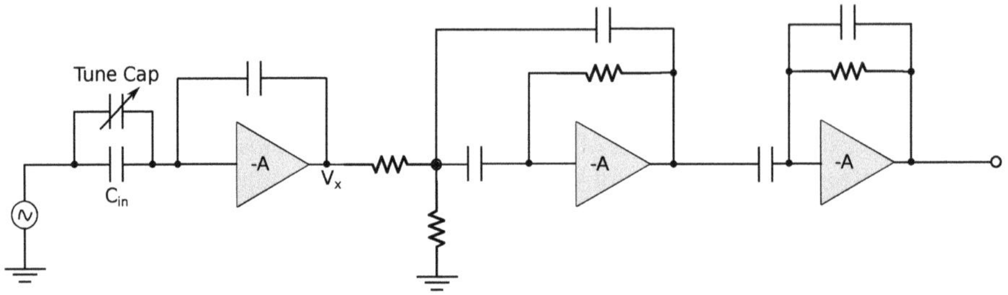

Figure 10: Simplified schematic of the 3-stage low-noise amplifier.

As core amplifier for the inverting voltage amplifier and CR-RC shaper a folded cascode topology (Figure 11) was implemented. This folded cascode is a single-ended amplifier with a PMOS as input transistor. The cascode transistors form an NMOS-PMOS structure. In addition, a load resistance is used to increase the voltage gain. Two outputs can be distinguished: one output refers directly to the cascode output and the second output is a source follower output. Generally, the idea of a cascode structure is to convert the input voltage to a current and feed the result to a common-gate stage [4]. The low-voltage path is activated through a select signal using the selection logic. The amplifier stages are designed to amplify signals starting from 100 kHz to 48 MHz at gain of 70 dB. A required frequency bandwidth between 500 kHz and 6 MHz is achieved [4].

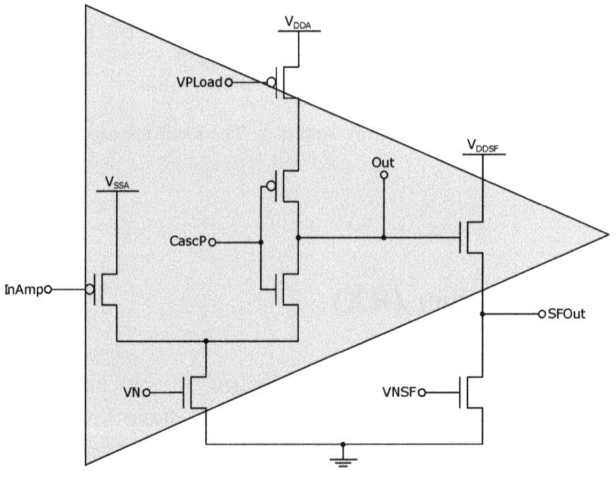

Figure 11: Folded cascode amplifier with PMOS as input transistor.

2.3 Digital Interface

To configure the ASIC an SPI (synchronous serial communication interface) was implemented as digital interface. A micro-controller is used to drive the SPI protocol. Further the micro-controller has a select line and a clock (SCLK) generated by the micro-controller. Generally, SPI devices communicate using a master-slave architecture. The SPI can select the ASIC using the select signal and acts as chip select. While data is being sent, the data written to the ASIC and is also read back. The clock is only enabled whilst these write/read operations to spare power consumption. To implement the digital interface it includes processes such as register transfer level (RTL) design, functional verification and synthesis. The backend process consists of floor planning, place and route, clock three synthesis, layout versus schematic (LVS) and GDS file generation.

3 Measurement Results

Recent measurements with prototype transducers are promising and fulfill the requirements of 3D USCT. The TX excitation, a linear chirp between 200 kHz to 4.7 MHz, was generated off-chip and amplified by the high voltage amplifier.

Figure 12: Measurement setup consisting of micro-controller and test board (yellow lines show the location of the ASIC bonded on the test board).

The test setup (see Figure 12) consisting of a micro-controller, test board where the ASIC is bonded and Qt test environment. The test board has several features to test all nine high-voltage and low-voltage outputs. For the measurement of the high-voltage amplifier an input

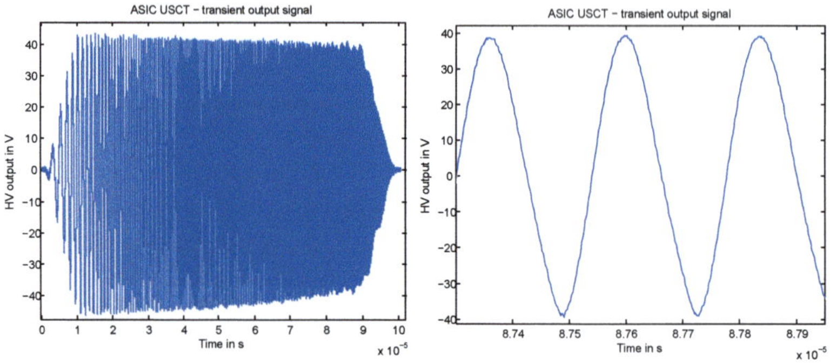

Figure 13: High-voltage output chirp signal (left). The output amplitude is 90.4 V (lower frequencies) and the measured gain is 18.4. Zoomed high-voltage amplitude (right) shows high linearity without distortions.

chirp signal (produced with a signal generator) with a sample rate of 40 MSa/s, amplitude of 4.9 V (peak-to-peak), offset of 1.4 V and an applied high-voltage of ± 48 V.

The measured high-voltage output is 90.4 V (Figure 13 left) at lower frequencies and the gain is 18.4. As expected high linearity (Figure 13 right) is given for low and high frequencies. The bandwidth between 200 kHz and 4.7 MHz is maintained. The Fast Fourier Transform (FFT) in Figure 14 shows the behavior of the output signal with respect to the input reference signal. The amplification is relatively constant with 18% drop in frequency range.

For the 3-stage low-noise amplifier measurements all low-voltage channels were activated. As explained in chapter 2.2 to obtain high linearity and a amplification of 3000 without oscillations of stage 3 a parameter set for tune capacitances and resistances was established. An input signal with the following parameters was used: rectangular waveform, frequency of 1 MHz, input amplitude of 1 V. The injection capacitance per channel on-chip was 1 fF. With the injection scheme the input signal of the amplifier input corresponds to 100 μV. The stage 3 produced an output amplitude up to 300 mV (peak-to-peak) leading to an amplification of 3000. Stage 3 shows the expected behavior. With the rising edge of the rectangular input the output of stage 3 rises and with the falling edge the output of stage falls accordingly. To measure the crosstalk of the channels between channels 0, 1, 2 one of the three channels is active while all others are off. Table 2 shows the results.

Channel (active)	Crosstalk (V/dB)
0 (560 mV)	2.9 mV / 45.7 dB (Channel 1)
0 (560 mV)	2.7 mV / 46.3 dB (Channel 2)
1 (545 mV)	3 mV / 45.2 dB (Channel 0)
1 (545 mV)	2.8 mV / 45.8 dB (Channel 2)

Table 2: Crosstalk measurement of the low-voltage channels.

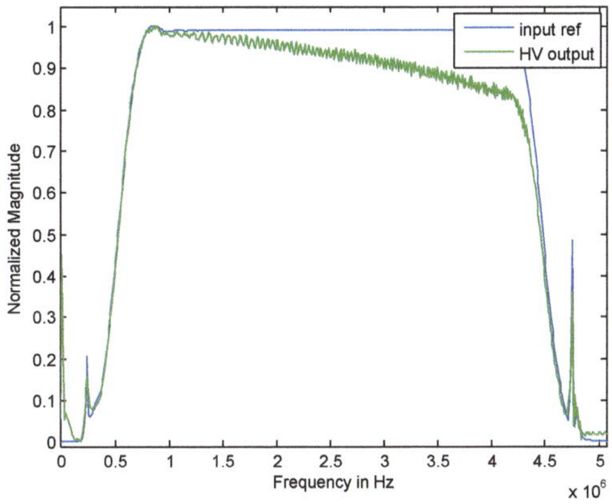

Figure 14: FFT of reference input chirp signal (blue line) and high-voltage chirp signal output (green line).

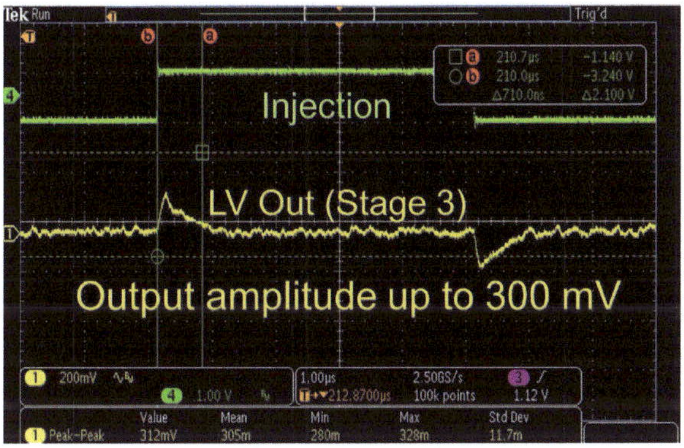

Figure 15: Injection input corresponds to 100 μV input signal. The calculated gain is 3000.

4 Conclusion

A new front-end electronic for USCT has been designed as an ASIC in 350 nm high-voltage CMOS technology. This ASIC is especially used for 3D USCT. The ASIC has nine channels consisting of high-voltage and low-noise amplifiers and a digital interface. The high-voltage transistors can sustain voltages up to 120 V. Simulations, design and measurements were presented in this paper. The ASIC enables with 18 combined receiver (RX) and transmitter (TX) elements. The performance of the high-voltage amplifier stage fulfills the USCT requirements. An output amplitude of 90.4 V and a bandwidth of 200 kHz to 4.7 MHz was achieved. Concerning the low-noise amplifier, a stable amplification up to 5000 was usable. A new version of this ASIC in 180 nm high-voltage CMOS technology is planned. This technology provides 200 V transistors. The same high-voltage circuit topology can be used. The number of channels in the new ASIC will be increased to augment the number of piezoelectric transducers.

References

[1] H. Gemmeke, L. Berger, T. Hopp, W.Y. Tan, I. Peric, M. Zapf, N.V. Ruiter: The New Generation of the KIT 3D USCT. Proc. International Workshop on Medical Ultrasound Tomography, 2017.

[2] Menshikov, A.: USCT Project Transduceer Array Head, Karlsruhe Institute of Technology, Institute for Data Processing and Electronics, 2008.

[3] H. Gemmeke, L. Berger, T. Hopp, M. Zapf, W. Tan, R. Blanco, R. Leys, I. Peric, and N. Ruiter: The New Generation of the KIT 3D USCT, Institute for Data Processing and Electronics. The 2nd International Workshop on Medical Ultrasound Tomography, 2017.

[4] R.Blanco: Customized Integrated Circuits for Scientific and Medical Applications, Karlsruhe Institute of Technology, Institute for Data Processing and Electronics, PhD Thesis, 2019.

Fast Auto-adaptive Gain Adaption for Improved Signal Dynamics

Z. Lu[1], M. Zapf[1], and N. V. Ruiter[1]

[1] *Karlsruhe Institute of Technology, Karlsruhe, Germany, E-mail: zewei.lu@kit.edu*

Abstract

In our 3D Ultrasound Computer Tomography system (USCT), the 12 bit ADC and factor 10 VGA are insufficient to resolve the smallest interesting signals. An adaptive front-end gain can solve this by object specific adaptions during the measurement.

The 3D USCT II of the KIT device contains 157 Transmitter Array System (TAS). Each TAS has 13 piezoelectric transducers, corresponding analog signal front end (AFE) and an MSP430FG66xx series microcontroller (MCU). All TAS are connected to a control board through a two-wire serial bus system.

Direct Memory Access (DMA) was used in the hardware to control the interrupt of the Universal Serial Communication Interfaces module (USCI). To complete the data transfer without occupying the MCUs of the TAS. A location-based general call was developed in the control system. The host transmits one frame long message to all TAS in a general call mode. This message contains the configurations of all TAS for the next measurement step. The address of each TAS corresponds to the location of each configuration in the long message. Thus, in the broadcast mode, each TAS only obtains the configuration information required by itself. With these two improvements, to configure all of the TAS can be reduced to less than 3 ms, which is the shortest measurement interval.

The here proposed solution allows a fast dynamic control of the front-end electronics during measurement without extending the measurement time significantly.

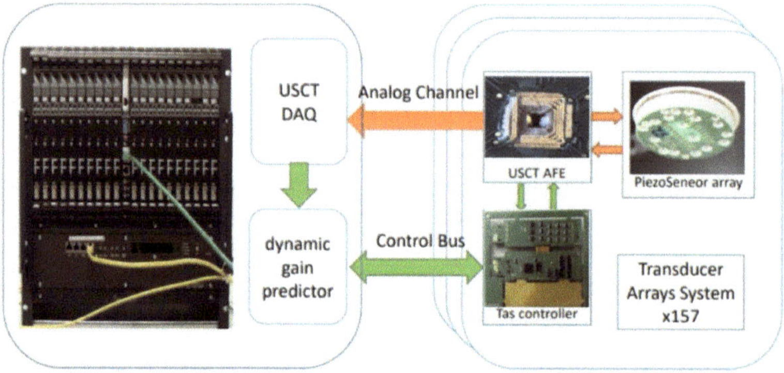

Figure 1: The figure shows the basic architecture of our system: combined with the USCT AFE, which is custom designed for USCT devices, the dynamic gain system can amplify the input signal in the range from 12 dB to 69 dB. Every channel has at least 5 MHz bandwidth.

First US Performance Measurements of Next Generation 3D USCT 2.5 Transducers

M. Zapf[1], M. Angerer[1], K. Hohlfeld[2], S. Gebhardt[2], H. Gemmeke[1], and N. V. Ruiter[1]

[1] *Karlsruhe Institute of Technology, Karlsruhe, Germany, E-mail: michael.zapf@kit.edu*
[2] *Technologies and Systems IKTS, Fraunhofer Institute for Ceramic, Dresden, Germany*

Abstract

The KIT's 3D Ultrasound Computer Tomography (USCT) II system has a multistatic setup of 2041 ultrasound transducers with approx. 1.5 MHz 6dB bandwidth and 36° 3 dB opening angle for 2.5 MHz. To increase the region of interest for a next USCT generation, the opening angle should be increased to approx. 60° and the bandwidth doubled. To increase the opening angle the size of the transducer elements was decreased to approximately half the size. A circular aperture was chosen for homogenicity of the radiation pattern in 3D. The transducer design utilizes piezo-fibres by the established Fraunhofer IMT piezo-fibre composite technology. The fibres were fabricated from PZT powder using the polysulfone spinning process. 17 fibres were positioned with a mechanical mask and filled with a matrix of epoxy. From this rod piezo composite discs were sawed and polarized. Electrodes were generated by silver-filled epoxy adhesive on the top and bottom side. Materials for acoustic backing is a Tungsten-Polyurethane composite and for acoustic matching ia aluminium oxide composite material (TMM4).

Ultrasound characteristics were evaluated quantitatively with a Onda HNC-400 hydrophone in a 3-axis water tank for a randomly selected sample transducer (see Fig. a.)). Characteristics evaluated were the pressure field as function over frequency and angle in the far-field (see Fig. b.)), following the use-case. For excitation a linear encoded chirp was used, for SNR improvements averaging of measurements (64 to 256 times) was conducted. The analysis compensated for the hydrophon's frequency and angular damping characteristics. The presented results show that the desired characteristics were mostly achieved: the 6 dB bandwidth could be vastly improved by roughly 200% (see Fig. d.)). The 6 dB pressure opening angle was approx. 50° (see Fig. c.)), not completly fullfilling the simulated expectations, an improvement by 31% was achieved. The results are promising for the next 3D USCT III generation.

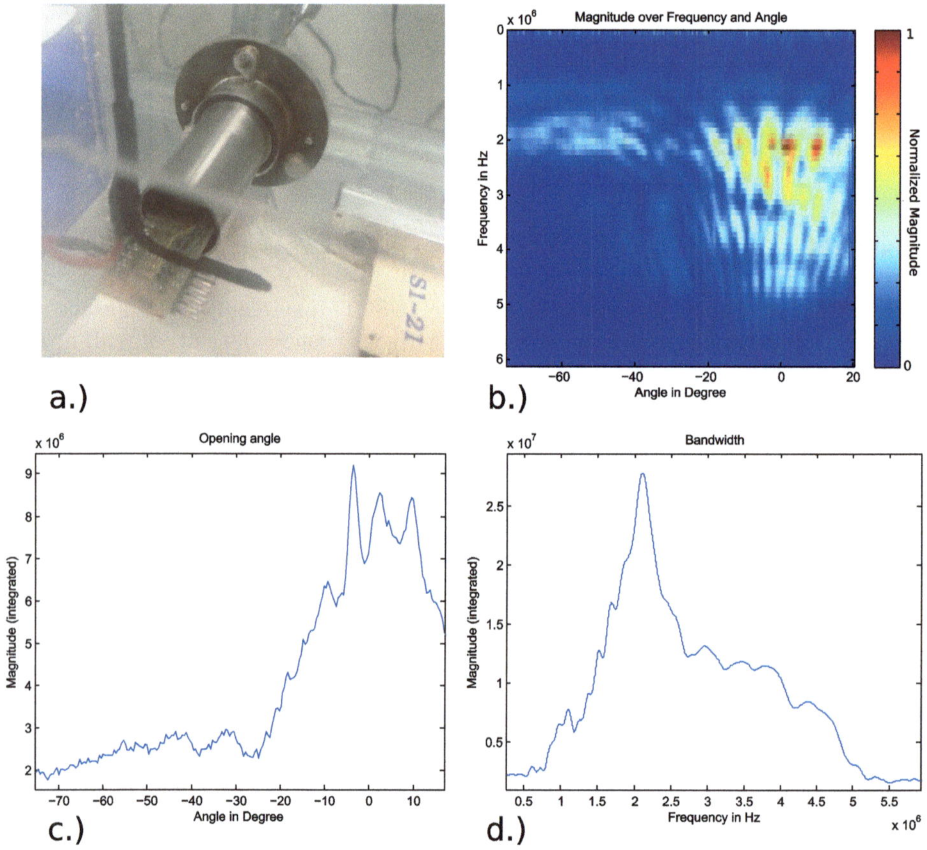
a.)
S1-21
b.)
Magnitude over Frequency and Angle
x 10^6
Frequency in Hz
Angle in Degree
Normalized Magnitude
c.)
Opening angle
x 10^6
Magnitude (integrated)
Angle in Degree
d.)
Bandwidth
x 10^7
Magnitude (integrated)
Frequency in Hz
x 10^6

Compensating for Variable Acoustic and Optical Properties towards Quantitative Photoacoustic Tomography

A. Pattyn[1], Z. Mumm[1], and M. Mehrmohammadi[1,2,3]

[1] *Wayne State University, Department of Biomedical Engineering, Detroit, USA*
[2] *Wayne State University, Department of Electrical and Computer Engineering, Detroit, USA*
[3] *Barbara Ann Karmanos Cancer Institute, Detroit, USA*

Abstract

Photoacoustic tomography (PAT) is a noninvasive, high-resolution imaging modality, capable of providing functional and molecular information of various pathologies such as cancer. In PAT imaging one optical limitation is the attenuation effect depth has on fluence and therefore the generation of PA signals from deeper regions of tissue. This makes it impossible to equally compare signals from various depths. In addition, most ultrasound and photoacoustic image reconstruction algorithms assume a homogeneous speed of sound distribution. This leads to a mismatch between the true location and size of the object and what the reconstructed image shows. This mismatch also plays a role in how effective and accurate fluence compensation can be. Since light diffusion is not affected by acoustic properties and the PAT image is, this leads to inaccuracies when fluence compensation is applied. In this study we investigate using a model-based method, that employs Monte Carlo light (MCXCL) and acoustic wave propagation (K-wave) simulations, to demonstrate the benefits of fluence compensation, in addition to the affect speed of sound correction has on fluence compensation. To generate our results we numerically modeled a full-ring illumination and ring-based acoustic detection system that our group has been developing. This model was then used to run an experiment that looked at a numerical phantom which was designed to test the affects of depth, concentration, and varying speed of sound on PAT image reconstruction. Our results indicate improvements in PAT optical fluence compensation and more importantly the feasibility of achieving qPAT images.

Design and performance of a Tonpilz transducer for low frequency medical ultrasound tomography

E. M. Lopes Filho[1], A. V. Pigatto[2], J. L. Mueller[2,3], and R. G. Lima[1]

[1] *Department of Mechanical Engineering, Polytechnic School of University of Sao Paulo, SP, Brazil, E-mail: ely.lopes@usp.br*
[2] *School of Biomedical Engineering, Colorado State University, Fort Collins, CO, USA*
[3] *Department of Mathematics, Colorado State University, Fort Collins, CO, USA*

Abstract

The design and performance of a Tonpilz transducer for low frequency ultrasound tomography is presented, motivated by recent research demonstrating that acoustic waves transmitting at frequencies between 10 kHz and 750 kHz penetrate the lungs and may be useful for imaging patients with COPD. An adaptation of the traditional Tonpilz design is presented, and the optimal frequency, directivity, and temperature sensitivity were analyzed through simulations and experiments. A static and dynamic analysis was also performed in simulations. The design is found to be suitable for the target application of pulmonary imaging with USCT between 50 and 200 kHz.

Keywords: Tonpilz, Ultrasound Tomography, Thorax Imaging

1 Introduction

Ultrasound has been used as a medical imaging diagnostic tool for more than 70 years. Most of the equipment relies on B-mode ultrasound, which generates sound waves of frequencies within the range of 2 to 10MHz. However, this frequency range poses a problem for pulmonary imaging, as the acoustic waves reflect off of the lung pleura, carrying no information about the interior of the lung and potential pathology. Recent research [1] has shown that there are three frequency bands in which acoustic signals behave in distinct ways in the human thorax. Between 1 and 10 kHz transmission is absent. Above 1 MHz, there is also no transmission through healthy lungs. However, there is an efficient frequency range for acoustic transmission between 10 and 750 kHz, with signal attenuation increasing as frequency increases. These properties pose an opportunity both for the diagnosis of COPD and bedside pulmonary monitoring in the ICU using low frequency ultrasound tomography. However, the low frequency signal, the geometry of the thorax, and the safety constraints of medical ultrasound pose a challenge in transducer design, and commercial transducers are not available for this application.

In this work, the Tonpilz transducer design, commonly used in sonar application, is adapted to have a more compact design and broader directivity for use in medical USCT. With the goal of tomographic thoracic imaging, an understanding of the transducer properties such as the temperature sensitivity, directivity, and acoustic power is necessary for obtaining optimal performance and assessing the safety of its use. Through simulations and experiments we demonstrate that the prototype meets both design and safety considerations for this application.

2 Methodology and Hardware Description

2.1 Transducer Design

The transducer development parameters were determined based on the following premises: the ability to operate at a frequency within the range of 20kHz to 750kHz, the external diameter must be 25 mm, and the thickness must not exceed 11 mm. The frequency limitation was based on the fact that sound waves of frequencies out of the proposed range cannot permeate the lungs [1]. The dimensional restrictions were imposed by the goal of thoracic USCT imaging with two rows of transducers placed circumferentially around the chest.

Tonpilz transducers are widely used in sonar applications due to their low cost, high acoustic power and simple structure [2]. A standard Tonpilz transducer is composed of a tail mass, a stack of piezoelectric ceramic rings (PZT) and a head mass that weighs less than half of the tail mass [3]. This topology is known for achieving a resonance frequency that no longer depends only on the PZT but on the masses of the parts used in its construction. Furthermore, lower working frequencies can be achieved than those found on commercial transducers, and thus, an adaptation of the traditional Tonpilz construction method was chosen for this project.

2.1.1 Simulation

A virtual model of the experimental transducer was developed and analyzed using the Finite Element Method (FEM). The 3D model can be seen on Figure 1. The transducer is composed of the tail mass, two PZT8 ceramics, separated by a thin copper sheet, and a piston; a silicone rubber o-ring is used to fill the gap between the piston and the body. The main constructive difference, when compared to the standard Tonpilz design, is that the tail mass is laterally distributed, which provides housing for the ceramics and piston. Furthermore, by spreading the mass laterally, the overall thickness can be reduced, which is desirable for thoracic imaging. During the design process, several characteristics of the transducer were simulated and analyzed, such as the resonance frequency modes, the directivity, and all the mechanical stresses occurring on each part. The final physical dimensions and construction materials were determined iteratively based on the application requirements.

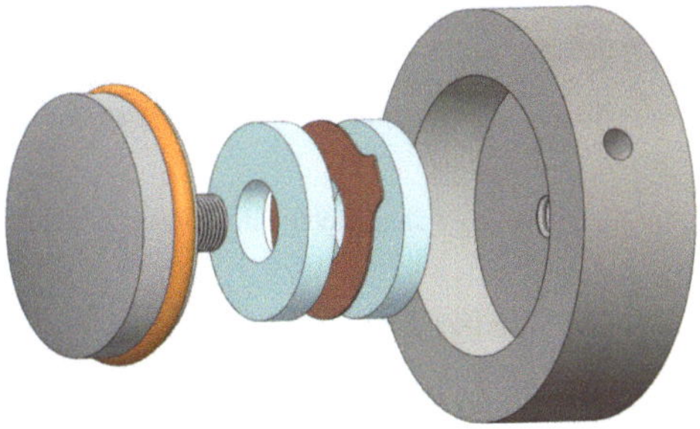

Figure 1: Virtual model of the experimental transducer.

2.1.2 Fabrication

After determining the optimal dimensions and materials through computational analysis, the transducer parts were fabricated with the aid of a milling machine with a tolerance of ±0.001". The tail mass was built using 304 stainless steel, while the piston was machined in 6061 T6 aluminum, increasing the difference of masses between those parts and ensuring that the highest fraction of the mechanical energy produced by the PZTs is transferred to the piston and not to the tail mass. Both the tail mass and piston are connected to the electric ground, while the internal copper sheet supplies the excitation signal to the PZTs, ensuring a safe operation. The cable that connects the transducer to the UST system was built using an RG-316 shielded cable and a male SMT terminal; moldable rubber protection was added to the transducer to ensure proper water sealing, as shown in Figure 2.

From Figure 2, it is noticeable that no bolts are attaching the piston to the tail mass since the threads were already machined on the piston's center pin. A torque of up to 0.3 N/m was used to assemble the parts, aiming to eliminate gaps between the PZTs, the copper contact, and the transducer body. After that, the sensitivity of all the transducers was tested for receiving and transmitting signals.

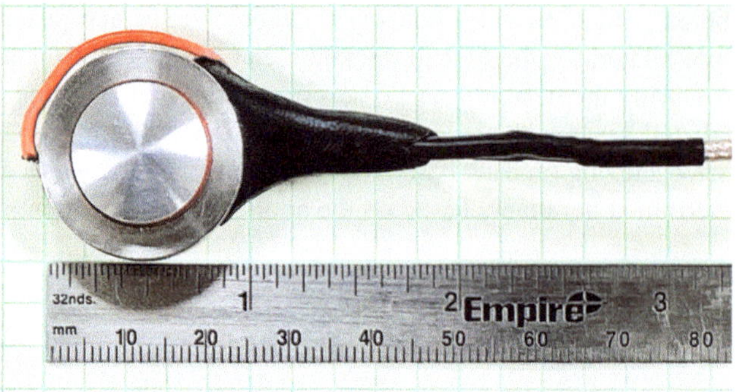

Figure 2: Final assembled transducer.

2.2 Experimental Protocol

A randomized block experimental design with n = 4 was performed to evaluate the transducer's behavior. The controlled factors were the water temperature (Factor A, 2 levels: 97.7°F and 104°F), the receiving transducer position (Factor B, 8 levels: 0° to 90° with a step of 11.25° referenced to the transmitting transducer position) and the transmission (Factor C, 2 levels: transmitting using transducer A and receiving with B and vice-versa). Figure 3 shows the configuration of the hardware used on the experiment.

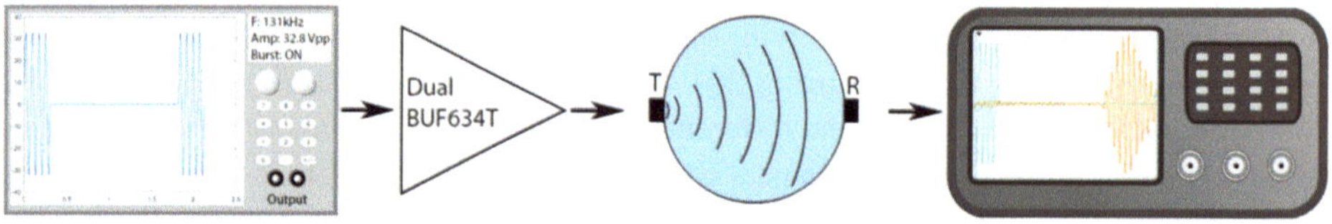

Figure 3: Experimental procedure hardware illustration.

As shown in Figure 3, a pair of transducers were positioned in a 274 mm diameter cylindrical tank, filled with water, and the position of the receiving/transmitting transducer was varied to cover each of the distinct angles and distances illustrated on Figure 4. An SRS DS360 Ultra Low Distortion Function generator was configured to produce two sine waves of 131 kHz, with an amplitude of 32.8 V_{pp} and a phase shift of 180°. The number of cycles per burst and repetition rate was adjusted to avoid interference among transmitted and reflected waves. Furthermore, the generated signals were used as input of a two-channel power amplifier assembled with two BUF634T high-speed buffers, and the outputs of the power amplifier were connected in a bridge configuration to excite the transmitting transducer with 23.2 V_{RMS}. A Tektronix TDS2001C oscilloscope was used to measure the amplitude and delay of the transmitted and received signals.

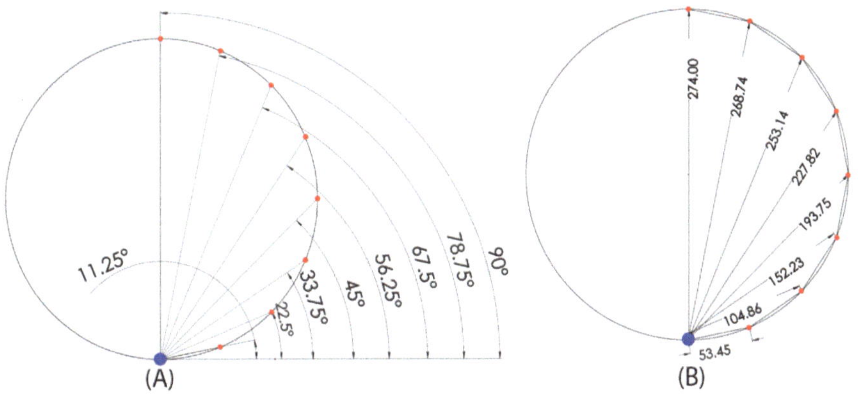

Figure 4: Illustration of the transducers position referenced to the origin (blue dot): (A) angles (B) distances.

The pair of transducers was positioned in eight different configurations, each of which with a particular angle and distance (see Figure 4). Therefore, a variation on the amplitude and phase delay between the transmitted and received waves was expected.

3 Results and Discussion

3.1 Simulation Results

The virtual model of the transducer was implemented with a tail mass of external diameter 25 mm, a thickness of 8 mm, and a piston of size 16 mm wide and 10 mm tall. The depth and width of the cavity on the tail mass were set to 5 mm and 17 mm, respectively. However, the piston head surface curvature was determined using FEM simulations, optimizing for the highest total irradiated acoustic power and a radiation pattern that suits the application; the results are shown in section 3.1.1. After the determination of the final dimensions that provide the desired acoustic properties, a mechanical analysis was performed to ensure the durability of the transducer; the results are shown in section 3.1.2.

3.1.1 Acoustic Data

In order to determine the piston head curvature resulting in the highest radiated power, heights from 0.5 mm to 1.5 mm were evaluated using a step size of 0.1 mm. After the first round of simulations, the height's range was shortened to 0.88 mm to 0.92 with a step of 0.005 mm; final results are shown in Figure 5.

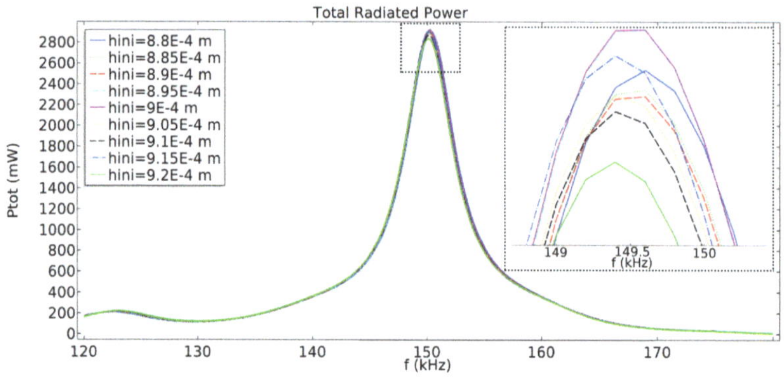

Figure 5: Maximum irradiated power for different piston head curvatures.

From Figure 5, one sees that the piston head curvature height that results in the highest total radiated power is 0.895 mm. Moreover, it is shown that the peak resonance frequency is at 149.5 kHz, and that the peak power is 2900 mW. The thermal index (TI) that can be achieved due to the frequency and the peak power is 2.06 [4], which is 67% below the maximum value approved by the FDA [5], and thus, considered safe. Therefore, the piston with a curvature of 0.895 mm was adopted as a standard for all the following simulations. Figure 6 illustrates the radiation pattern and the acoustic pressure map of the transducer when excited with a sinusoidal source with an amplitude of 30 VRMS and a frequency of 149.5 kHz.

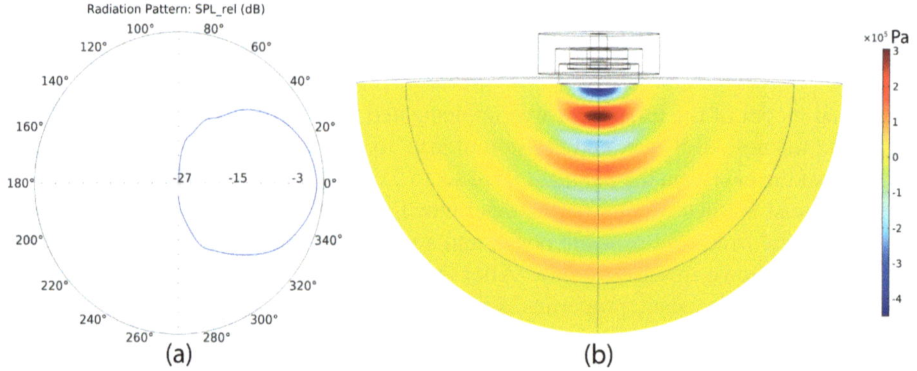

Figure 6: (a) radiation pattern, (b) acoustic pressure map.

As seen in Figure 6 (a), the beam divergence, angle from the acoustic axis to the point where the acoustic pressure has decreased by one half (-6bB), is 38°. The maximum acoustic pressure achieved was found to be 0.345 MPa, which results in a mechanical index (MI) of 0.89 [6]. The FDA approved maximum MI for diagnostic imaging is 1.9 [5].

3.1.2 Structural Analysis

To analyze the behavior of the structure of the transducer, a static and dynamic stress analysis was conducted. The Von Mises stress was determined on the condition where the maximum voltage is applied to the transducer and a torque of 0.3 N/m was used during its assemble; the results are shown on Figure 7.

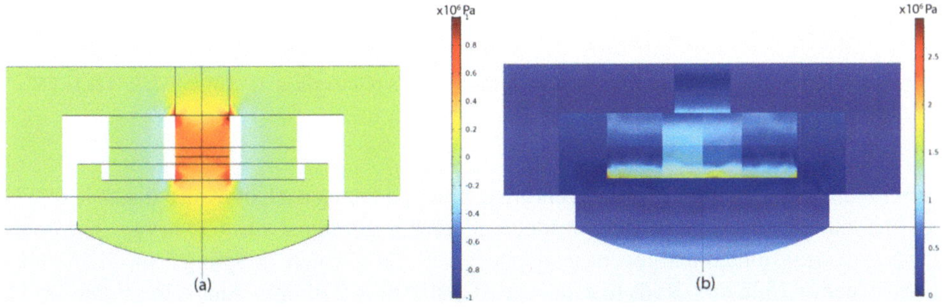

Figure 7: (a) static mechanical stress, (b) dynamic mechanical stress.

As shown in Figure 7 (a), the maximum static stress region occurs on the piston bolt and reaches a value of 100 MPa. The static stress is caused by the force necessary to compress the PZTs and internal parts of the transducer and is solely dependent on the manual torque applied to the piston during assembly. From Figure 7 (b), it is noticeable that the maximum stress caused by the vibration of the PZTs occurs on the back surface of the piston and reaches a value of 2.8 MPa. Therefore, considering both sources of deformation combined, a safety factor of 2.33 is achieved since the yield strength of the material of which the piston was designed is 240 MPa. Furthermore, different tightening torques were simulated, and the results have shown that an increase in torque over 0.3 N/m did not improve the transducer acoustic characteristics.

3.2 Experimental Results

The main purposes of the experiment were to evaluate the acoustic radiation pattern of the transducer and the effect of the temperature variation on its sensitivity considering the human body temperature range. Therefore, two variables were measured and analyzed – the voltage on both transmitting and receiving transducers and the time of flight (T.o.F), which is the time delay between the transmitted and received signals [7]. This procedure was repeated four times, for all of the angles shown in Figure 4, for two different temperatures. The position and order of measurement were randomized, while the temperature was fixed. Figure 8 presents the voltage signals measured with the oscilloscope for three different angles but kept under the same temperature.

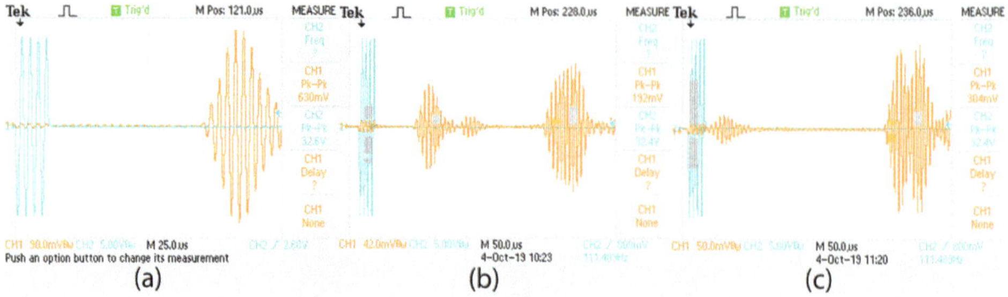

Figure 8: Voltage measured on the receiving transducer when it is positioned at: (a) 90° (b) 33.75° (c) 11.25°.

From Figure 8 (a), where the transducers are positioned facing each other at a distance of 274 mm, the voltage measured on the transducer receiving the signal was 630 mV_{pp}. The delay between the transmitted and received signals is 177.5 μs, which, adopting a speed of sound on the water of 1522 m/s, leads to a calculated distance of 270 mm. Analyzing Figure 8 (b), it is seen that the voltage of the received signal was 130.2 mV_{pp} and the T.o.F is 105 μs, thus, the calculated distance is 159 mm. Furthermore, it is interesting to point out that the amplitude of the second received signal is higher than the first one. One possible explanation is that wave reflections on the tank may be interacting with the receiving transducer on an angle to which it is more sensitive than the one from where the primary wave is coming from. The same behavior is seen in Figure 8 (c), where the T.o.F is 38 μs, the calculated distance is 57 mm and the amplitude of the received signal is 56 mV_{pp} whereas the amplitude of the second lobe is 304 mV_{pp}.

3.3 Statistical Analysis

The data collected during the experiment was analyzed using the Minitab 18 stats tool, where an analysis of variance was conducted with a confidence interval of 99% (F-Table a = 0.01). The residuals showed a normal distribution, and the equal variances premise was not violated, and thus, it was considered that ANOVA is an adequate test for this experiment [8]. ANOVA showed that the only controlled factors that significantly affect the received signal amplitude are the transducer position on the cylindrical tank (F-Value = 35.13 P-Value = 0.000) and which transducer is transmitting/receiving the signal (F-Value = 13.2 P-Value = 0.000). The temperature did not significantly affect the response variable (F-Value = 0.51 P-Value = 0.478). Figure 9 shows the average amplitude of the received signals as a function of the receiving transducer position referenced in polar coordinates, as adopted in Figure 4.

From Figure 9, a clear rise on the average amplitude of the received signal in function of an increase on the angle can be seen. At the position referenced by 90°, the amplitude of the received signal was 664.4 $mV_{pp} \pm$ 57.9 mV_{pp}, whereas at 11.25°it was 56.7 mV_{pp} $\pm$ 8.71 mV_{pp}, which represents an attenuation of 21.4 dB. Thus, it was considered that

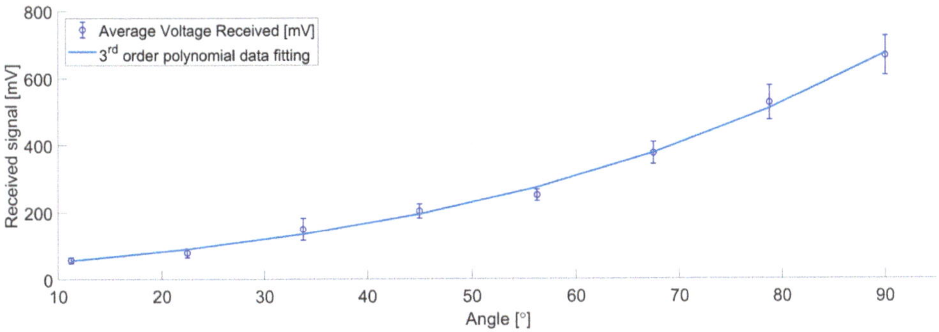

Figure 9: Voltage measured on the receiving transducer as a function of its position.

these results were compatible with the radiation pattern simulated and shown on Figure 6. Analyzing the data acquired a clear difference in the amplitude of the signal received when transducer B was used as the transmitting source was noticed. For the same conditions (position and temperature), the magnitude of the voltage measured on transducer A, when used as a receiver, was $7.8 \pm 1.1\%$ higher than when transducer B was used as a receiver. As the difference was consistent, i.e., always positive and within a range of relative variability of 14was attributed to the fact that the sensitivity of the transducers to convert the electrical signal in mechanical and vice versa might be significantly different. Several factors might have caused the variability on the sensitivity of the transducer on the experiment, such as the tolerances of the piston and tail masses and PZT8 parameters, the crimping position of the sensor in the cylinder among measurements and even the preload given by the torque applied to the piston. Previous studies have shown a current and voltage amplitude gain on a transducer built using PZT8 crystals when the preload was increased from 44 MPa to 96 MPa [9]. A few other studies that analyze the behavior of PZTs under high temperature have been conducted with a single PZT-4 or PZT-8 ceramic to avoid the variability among samples [10, 11]. Therefore, it is clear that a sensitivity variance is expected for the transducer, and thus, a calibration is required.

4 Conclusion

This study aimed to develop a Tonpilz transducer capable of operating on a low frequency range (less than 750 MHz) for the application of pulmonary USCT. The maximum acceptable external dimensions of the transducer were determined, and several simulations were conducted to analyze its behavior. Through the acoustic analysis, it was shown that the maximum acoustic pressure achieved is safe for use on human patients and that the beam divergence is adequate for the application. The structural analysis results have shown that the safety factor is 2.33 and that the preload torque is responsible for the primary stress. From experimental data collected on a water-filled tank, the effect of the position of the receiving transducer on the amplitude and delay of the received signal and the effect of the temperature over the sensitivity of both transducers were measured and analyzed. Through the statistical analysis, it was found that variations in temperature in the range of those of the human body do not significantly affect the transducer sensitivity. However, there were significant differences when the receiving and transmitting transducers were switched, which indicates that a calibration will be necessary to ensure measurement accuracy. Therefore, as a preliminary study, it was considered that promising results were achieved.

Acknowledgement

The project was supported by Award Number R21EB024683 from the National Institute of Biomedical Imaging and Bioengineering. The content is solely the responsibility of the authors and does not necessarily represent the official view of the National Institute of Biomedical Imaging and Bioengineering or the National Institutes of Health.

References

[1] D. Rueter, H. P. Hauber, D. Droeman, P. Zabel, S. Uhlig, Low-frequency ultrasound permeates the human thorax and lung: A novel approach to non-invasive monitoring, Ultraschall in der Medizin 31 (1) (2010) 53–62.

[2] Q. Yao, L. Bjørnø, Broadband tonpilz underwater acoustic transducers based on multimode optimization, IEEE Transactions on Ultrasonics, Ferroelectrics, and Frequency Control.

[3] R. J. Meyer, T. C. Montgomery, W. J. Hughes, Tonpilz transducers designed using single crystal piezoelectrics, in: Oceans Conference Record (IEEE), 2002.

[4] H. C. Starritt, F. A. Duck, Safety, in: Clinical Ultrasound, Churchill Livingstone, 2011, pp. 51–60.

[5] FDA, Marketing Clearance of Diagnostic Ultrasound Systems and Transducers; Guidance for Industry and Food and Drug Administration Staff, U.S. Department of Health and Human Services: Food and Drug Administration: Center for Devices and Radiological Health (2019) 18–34.

[6] R. E. Apfel, C. K. Holland, Gauging the likelihood of cavitation from short-pulse, lowduty cycle diagnostic ultrasound, Ultrasound in Medicine and Biology.

[7] D. Marioli, C. Narduzzi, C. Offelli, D. Petri, E. Sardini, A. Taroni, Digital Time-of-Flight Measurement for Ultrasonic Sensors, IEEE Transactions on Instrumentation and Measurement.

[8] D. C. Montgomery, Design and Analysis of Experiments, 8th Edition, Vol. 1, Wiley, Hoboken, 2012.

[9] D. A. Deangelis, G. W. Schulze, Performance of PZT8 Versus PZT4 Piezoceramic Materials in Ultrasonic Transducers, in: Physics Procedia, 2016.
[10] L. Tang, W. Cao, Temperature dependence of self-consistent full matrix material constants of lead zirconate titanate ceramics, Applied Physics Letters 106 (5).
[11] Y. Zhang, L. Tang, H. Tian, J. Wang, W. Cao, Z. Zhang, Determination of temperature dependence of full matrix material constants of PZT-8 piezoceramics using only one sample, Journal of Alloys and Compounds 714 (2017) 20–25.